Telephone Repair Illustrated

Telephone Repair Illustrated

Stephen J. Bigelow

TAB Books
Imprint of McGraw-Hill

New York San Francisco Washington, D.C. Auckland Bogotá
Caracas Lisbon London Madrid Mexico City Milan
Montreal New Delhi San Juan Singapore
Sydney Tokyo Toronto

To my darling wife Kathleen.
Without your loving encouragement and support,
this book would still have been possible—
but not nearly worth the trouble.

Notices
Touch-Tone™   AT&T
AMPS™   AT&T
Kapton®   DuPont de Nemours Co.

pbk   6 7 8 9 10 11 12 13 14 FGR/FGR 9 9 8 7
hc   9 10 11 12 13 14 15 16 17 18 FGR/FGR 9 9 8 7 6

Library of Congress Cataloging-in-Publication Data

Bigelow, Stephen J.
    Telephone repair illustrated / by Stephen J. Bigelow.
        p.   cm.
    Includes index.
    ISBN 0-8306-4034-7 (pbk.)   ISBN 0-8306-4033-9 (hard)
    1. Telephone—Maintenance and repair—Amateurs' manuals.
I. Title.
TK9951.B54 1993
621.386—dc20   92-32381
                                                      CIP

Acquisitions editor: Roland S. Phelps
Editorial team: Anita Louise McCormick, Editor
                Lori Flaherty, Supervising Editor
                Joanne Slike, Executive Editor
Production team: Katherine G. Brown, Director of Production
                Rhonda E. Baker, Layout
                Ollie Harmon, Typesetting
                Tina M. Sourbier, Typesetting
                Tara Ernst, Proofreading
Design team: Jaclyn J. Boone, Designer
                Brian Allison, Associate Designer
Cover design: Graphics Plus, Hanover, Pa.   EL1
Cover photograph: Brent Blair   4179

Contents

Acknowledgments

I WOULD LIKE TO THANK THE FOLLOWING INDIVIDUALS AND ORGANIZATIONS for their generous contributions of photographs and technical information used throughout this book:

Mr. Tony Magoulas, Marketing Information Representative for Radio Shack (a division of Tandy Corporation). Mr. Gregg Elmore, Manager, Marketing Communications for B+K Precision. Ms. Jan Smith, Director of Marketing for Code-A-Phone Corporation.

Introduction

IT'S HARD TO BELIEVE THAT TELEPHONES HAVE BEEN AROUND SINCE 1876 WHEN Alexander Graham Bell uttered his immortal words, "Mr. Watson! Come here, I want you!" Bell's telephone instrument gained almost immediate acceptance as a means of personal and business communication over long distances. Supported by a growing national network, it did not take long for the telephone to find its way into homes and businesses across America.

Ultimately, the telephone network has grown to encircle the world, and Bell's simple invention has become an indispensable part of our everyday lives. The advent of sophisticated electronics has allowed a wide variety of "intelligent" features and enhancements to be added, making the ordinary telephone even more valuable. Yet, in spite of their inherent reliability (we all tend to take telephones for granted), today's complex telephones, like most other electronic instruments, do fail from time to time. When failures occur, you can often find and repair the trouble yourself.

This book is written with two purposes in mind: first, it was written to teach you the technologies behind telephones—what they are, how they are built, how they work, etc.; second, it is a guide for troubleshooting and repair. The book covers a wide variety of telephone equipment—basic electromechanical telephones, electronic telephones, cordless telephones, and even cellular telephones. A chapter on answering machines is included as well.

The book is written primarily for the beginning to intermediate electronics enthusiast interested in learning about telephones and their repair. I have tried to prepare a book that is comprehensive enough to deal with a broad base of equipment, yet be specific enough to offer clear, hands-on guidance in troubleshooting and repair.

Your comments and suggestions about this book are welcome, as well as any personal telephone troubleshooting experiences you might wish to share. Feel free to write to me directly. I look forward to hearing from you.

Stephen J. Bigelow
Dynamic Learning Systems
P.O. Box 805
Marlboro, MA 01752

1
Telephone technology

THE PUBLIC TELEPHONE SYSTEM HAS SOMETIMES BEEN REFERRED TO AS THE eighth wonder of the modern world. Consider a system that is controlled by the world's most sophisticated network of interconnected computers, a system with the ability to connect any two telephones anywhere in the world—typically within just a few seconds. With more than 100 million telephones in the United States alone, you can appreciate the power of Alexander Graham Bell's "grand system." Not only can today's telephone network handle high-quality voice communication among individuals, it can also service the demands of modern, high-performance data transmission between computers.

Yet, in the face of this remarkable complexity, telephone instruments themselves (Fig. 1-1) are surprisingly simple devices that can be operated successfully by children only four or five years old.

Before you dive right into the components and operations of the telephone, it helps to get an overview of the Public Switched Telephone Network (PSTN), the general elements of every telephone, and the characteristics of signalling and transmission found in your ordinary telephone line.

The PSTN

The Public Switched Telephone Network is organized somewhat like a grove of trees whose roots are tied together. While this might be rather difficult to visualize at first, consider the diagram of Fig. 1-2. Although the figure is simplified, it illustrates the network's basic structure.

1

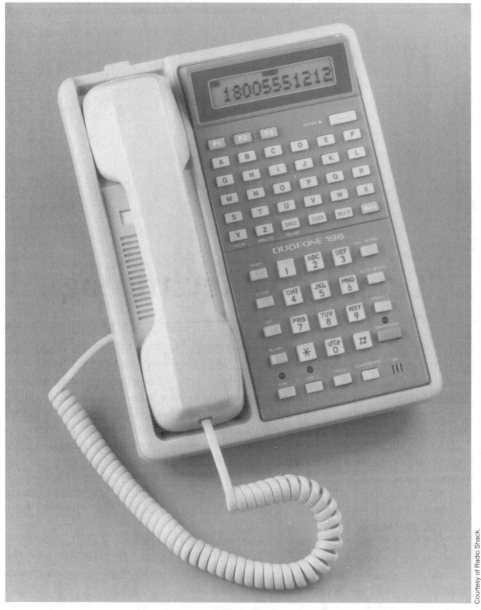

1-1   A typical "intelligent" telephone.

Courtesy of Radio Shack.

The central office

Telephones represent the very endmost points in the network. Each telephone is connected, usually by a pair of copper wires, to a local switching center known as a *central office*, or CO. A CO handles all switching functions between its local telephones, so if a call is made from one local telephone to another that is serviced by the

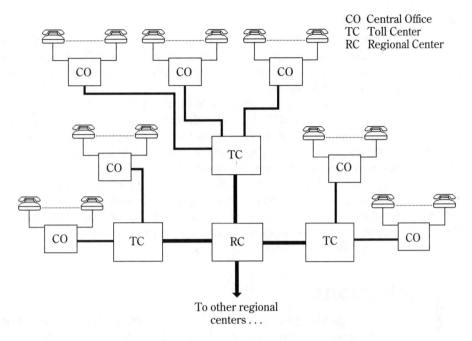

1-2   The PSTN.

same CO, the call will be switched and carried by the same CO without any further interaction with the network. This is often referred to as a *local call.*

Each CO is specified by a unique three-digit number such as 276 or 863—the first three digits of your seven-digit telephone number. The remaining four digits of your telephone number represent the specific telephone serviced by that central office. For example, if you dial the number 555–1212, then you are trying to reach telephone 1212 in central office number 555. Using this type of numbering scheme, a single central office can accommodate 10,000 telephones (xxx–0000 to xxx–9999).

The toll center

When you dial a telephone number that is serviced by a central office other than the one handling your own telephone, both central offices must be connected. Switching control from the originating CO is transferred to a *toll center,* or TC. The TC establishes the link between the originating and destination CO then transfers switching control to the destination CO. From there, the destination CO switches the call to the desired telephone, then signals the desired telephone that a call is waiting. If the desired CO is not available from the current toll center, the call must be switched to the next higher level in the network.

Although they are called toll centers, a charge is not always applied to the call—it really depends on just where the destination CO is located. For highly populated or urban areas that require several COs in the same general area, calls placed between adjacent COs are usually allowed without any toll charge. Where COs are separated by great physical distances, however, toll charges are often applied.

The regional center

When a call cannot be routed by a single TC, it is handed off to a *regional center,* or RC. At this level of the network, the call is routed to the necessary TC that handles the desired CO. From there, the call is switched normally at the desired CO and connected to the destination telephone.

If the RC receiving the call is not connected to the necessary TC, the regional center will switch the call to another RC that does handle the necessary TC. Using this type of hierarchy, it is possible to search out and connect any two telephones. Of course, the farther apart both telephones are, the greater the required hierarchy needed to connect them. This ties up more equipment during your call—and is charged accordingly.

Realistically, the actual telephone network is much more involved than the illustration shown in Fig. 1-2. Centers and offices are connected using copper wire cables, land- and space-based satellite microwave links, and high-capacity fiber-optic cables. The actual complexity of the network at any given point in the world depends on the terrain in the region and the demand for services.

The telephone

A telephone forms the beginning and end points of the PSTN. It is the telephone that allows us to interface with this vast network. While today's electronic telephones offer many more features than even a decade ago, every telephone must perform at least seven distinct functions. It must:

1. Request the use of the network from its local CO.
2. Inform you of the network status. This is normally done using selected combinations of tones.
3. Inform the CO of the desired number.
4. Inform you when a call is incoming.
5. Release its use of the network when the call is complete.
6. Transmit your speech onto the network, and receive the speech of a distant caller from the network.
7. Perform all these functions under an incredible variety of power levels and telephone line lengths.

Through the years, telephones have evolved to handle each of these functions efficiently and inexpensively. The block diagram of Fig. 1-3 illustrates the breakdown of a typical telephone. Take a moment to become familiar with each function—I refer to them throughout the book.

Tip and ring

No matter how "intelligent" or sophisticated your telephone is, its physical interface to the PSTN takes place through only two copper wires. These wires are known as *tip* and *ring*. Telephone wiring conventions specify the tip as a green wire, and the ring as a red wire. These names date back to the early days of telephones when all central office switching was performed by human operators who manually plugged and unplugged jacks to complete the telephone-to-telephone connections.

Tip (green wire)

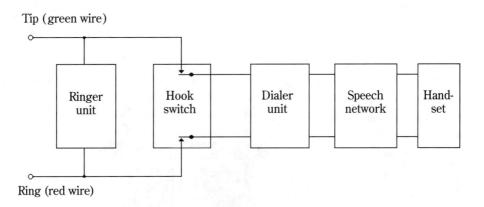

Ring (red wire)

1-3   A basic telephone unit.

The jacks into which the plugs were placed had two contacts: the tip of the jack made contact with the ball-like tip on the phone plug and the ring of the jack made contact with the barrel-shaped common portion of the plug. Of course, these terms have little practical meaning in modern telephony, but they are still used and are used throughout this book.

The ringer

A ringer is a device used to alert you to an incoming call. Technically, a ringer can be any device (audible or visual) that attracts your attention. Electromechanical ringers, such as the dual-bell ringer shown in Fig. 1-4, was originally patented by Thomas A. Watson (Bell's assistant) in 1878, just two short years after the telephone's invention.

When the central office signals an idle telephone, it sends out brief bursts of between 90 to 120 Vac at a frequency of about 20 Hz. This alternating current (ac) signal turns the ringer coil(s) on and off alternately. Magnetic forces produced in the coil(s) will rock a pivoted arm back and forth. This, in turn, swings a metal clapper that strikes one or two metal gongs. It is the physical size and shape of these gongs that ultimately determines the ringer's sound.

Ringing signals are produced in bursts. The pattern of these ringing bursts is known as *ringing cadence*. In the United States, ringing cadence is usually broken into 2 seconds of ringing, and 4 seconds of silence. Other countries might use a different ringing cadence. For example, the United Kingdom's ringing cadence consists of two short bursts of 0.4 seconds each, separated by a pause of 0.2 seconds. This pattern is broken up by a 2-second silence. Figure 1-5 illustrates a comparison between U.S. and U.K. ringing cadence.

Notice in Fig. 1-4 that a capacitor is added in series with the ringer coil(s). This capacitor (often about 0.1 μF) is included to block the flow of dc through the ringer at all times but allow ac ringing signals to pass. Without this capacitor, low coil resistance would cause the telephone to draw loop current just as if it were offhook.

Today, electromechanical ringers have essentially been replaced by electronic ringers based on integrated circuits. Electronic ringers are more popular because

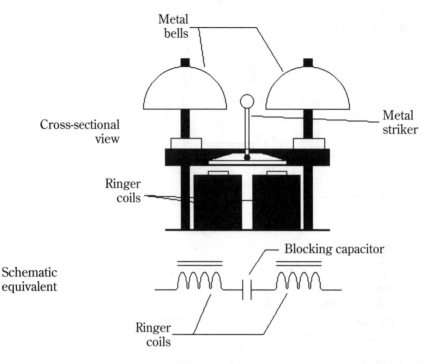

1-4   An electromechanical ringer.

they are small, light, easier (and less expensive) to manufacture, and they produce a more pleasing sound than the obnoxious clanging of an electromechanical ringer. The circuit fragment in Fig. 1-6 represents just one of the many electronic ringer ICs currently available. The actual sound is produced by a piezoelectric buzzer.

The hook switch

A hook switch is just that—a switch. It is a set of electrical contacts that connect (or disconnect) the telephone's voice circuit from the PSTN. Older telephones use a

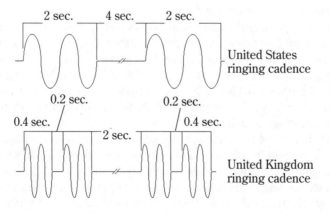

1-5   Typical ringing cadence patterns.

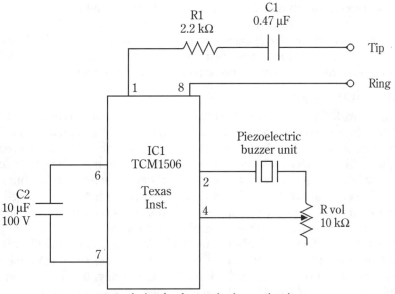

1-6 A simple electronic ringer circuit.

hook switch assembly with multiple sets of contacts that are actuated by the weight of the handset resting on the telephone. When the handset is resting in its "cradle," the hook switch contacts are opened and the telephone is effectively disconnected (or idle). When the handset is lifted from its cradle, spring tension closes the hook switch's contacts that connect the telephone to the PSTN.

In today's electronic telephones, these mechanical contacts have largely been replaced by a discrete on/off switch that activates a sealed relay. The relay contains similar contacts to mechanical contacts, but it can be controlled digitally. Cordless and cellular telephones usually use a relay technique.

The handset

Traditionally, the handset assembly contains a telephone's transmitter and receiver, which are connected into the telephone's speech network through a long spiral cable.

A transmitter converts sound vibrations from your voice into a varying electrical current that is then placed onto the network. Classical telephones use a carbon microphone to convert speech into energy. The carbon microphone is essentially a rigid metal diaphragm mounted over a sealed capsule containing packed carbon granules. An electrical current is passed through the carbon capsule, which now functions as a resistor. As sound energy strikes the metal diaphragm, the capsule expands and compresses. This changes the density of the packed carbon filling and results in a resistor that varies with sound energy. These resistance variations cause speech current to vary—the resulting electrical signal represents speech. Most electronic telephones now use an electret or electrodynamic microphone in place of the traditional carbon microphone.

A receiver reverses the process by converting speech signal current back into sound waves you can hear. A basic receiver consists of a rigid metal diaphragm placed closely to a permanent magnet wrapped with a coil of wire. When voice signal current is passed through the coil, the resulting electromagnetic field interacts with the permanent magnet, which causes the diaphragm to vibrate. This vibration creates sound that re-creates the speaker's voice. While the principles of telephone receivers have changed little since their original invention, better materials and designs have resulted in smaller, more reliable receiving elements.

The speech network

A speech network also called a *voice network,* or a *voice circuit,* serves a variety of important functions in your telephone. First, it converts the four signal-carrying wires from the handset (two wires for the transmitter, and two wires for the receiver) into the two-wire (tip and ring) telephone line—referred to as the *hybrid* function. The speech network also interfaces signals from the dialer to the two-wire telephone line. Finally, the speech network supplies automatic compensation for line length variations that keep the volume level of speech at a constant level. Electronic telephones also have amplifiers to augment the transmitter and receiver.

While older mechanical telephones used a bulky transformer and discrete components to construct a speech network, electronic telephones can often implement a complete speech network on a single IC, including filtering, line length compensation, and the four-wire to two-wire interface used to require a transformer.

The dialer

Whenever you pick up your telephone to initiate a call, your local CO must be signalled as to exactly which number (which destination telephone) is being called. This signalling is handled by a dialing unit. There are three common types of dialing used today: rotary, dual tone multifrequency, and pulse.

A rotary dialer, as shown in Fig. 1-7, has been used for many years, and can still be found on many older classical telephones. It consists of a finger plate connected to an intricate series of springs, cams, and governers that operate two sets of electrical contacts, as shown in Fig. 1-8. Whenever the finger plate is moved from its rest position, one of the contact sets closes. This mutes the sound passed to your receiver, and prevents annoying dialing "clicks" from being heard. Once the finger plate is released, carefully governed spring forces will pull the plate back to its rest position.

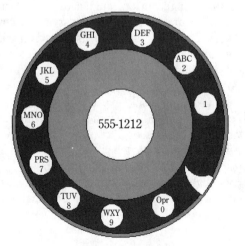

1-7 A mechanical rotary dial unit.

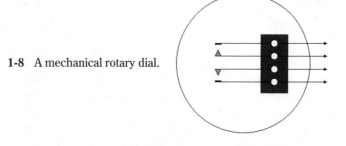

1-8 A mechanical rotary dial.

As the finger plate rotates, a mechanical cam opens and closes the second set of contacts, which interrupts the loop current flowing through your telephone. Each number dialed has a corresponding number of "clicks," so if you dial the number 3, the loop current will be interrupted three times as the finger plate moves to its rest position, dialing the number 7 produces seven clicks, and so on. The CO receives these clicks, and interprets them as the desired telephone number. You will learn more details of the rotary dial's operation in chapter 6.

The demands for greater dialing speed, combined with the early use of transistors in telephones, gave rise to a new form of dialing. Instead of sending out long strings of current pulses, desired digits are represented by unique combinations of audio tones. This technique is called dual-tone multifrequency) dialing, or DTMF. This type of dialing often goes by the name *Touch Tone*. The layout of a typical DTMF keypad, along with frequencies for each row and column, is shown in Fig. 1-9. For example, if the number 2 were pressed on your DTMF keypad, a 697-Hz and a 1336-Hz tone would be mixed together and delivered to the two-wire telephone line.

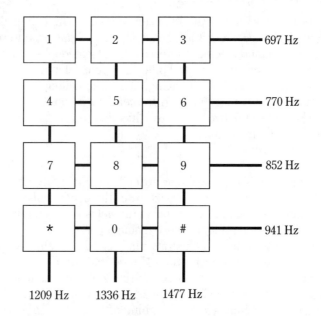

1-9 A standard DTMF keypad layout and frequency pairs.

Older-model DTMF keypads used individual transistor oscillators—one for each required frequency. Each oscillator was tuned with an inductor and capacitor (known as an *LC oscillator*). To handle audio frequencies, the inductors used in many early keypads were bulky and heavy. The inductors could also drift because of their age or variations of temperature and humidity in the environment. This often resulted in a shift of frequencies. Gradually, discrete LC oscillators were replaced with a single-tone generator IC using a piezoelectric crystal to set a reference frequency. All other frequencies could then be derived from the reference. The tone pairs are selected directly from the keypad's switch inputs. Electronic telephones often use ICs that incorporate the dialing function with other telephone features. A large selection of telephone ICs are presented throughout the rest of this book.

Once integrated circuits became widely accepted in telephone designs, it did not take long to discover that ICs could be made to simulate the opening and closing of rotary dial contacts. This led to a hybrid dial IC called the *IC pulse dialer* (also known as the *tele pulse dialer*. Instead of tones being generated when a key is pressed, the IC generates "clicks" that interrupt the telephone line just as a rotary dialer does. IC pulse dialers proved useful for people who wanted the convenience of pushbutton dialing but did not have DTMF facilities available at the local CO.

Because you can dial a number on a pushbutton keypad much faster than a digit can be pulsed, it was necessary to store keyed numbers until each could be sent. This rudimentary memory capacity evolved into the "last number redial" and multi-number memories that are commonplace in most electronic telephones today. Most IC dialers can now provide both DTMF and pulse dialing from the same IC—the specific dialing mode can be selected by a switch.

Telephone signalling

Signalling refers to the various methods used to control the connections between calling or called telephones. Signalling can also be the means used to report the status of a call as it is being connected. Generally, the "local loop"—that portion of the telephone network between your central office and your telephone—uses two types of signalling: dc signalling and ac signalling. Digital signalling is gradually gaining acceptance as a third type of signalling. A tremendous amount of signalling occurs throughout the rest of the network, but this has little direct impact on the performance of your particular telephone.

Dc signals

Dc signalling is based on the presence or absence of current in the local loop. This type of signalling is used to request service from the central office once your telephone is offhook (active), or release its use of the network when it is onhook (idle). Dc signalling is also used in rotary or pulse dialing operations.

An idle (onhook) telephone is illustrated in Fig. 1-10. The hook switch is open, preventing any loop current from flowing. Notice that the CO provides a −48-Vdc (measured from tip to ring) power source to drive loop current and carry voice signals. This is known as *battery* and can be measured with an ordinary multimeter.

When a telephone becomes active by going offhook, the hook switch contacts

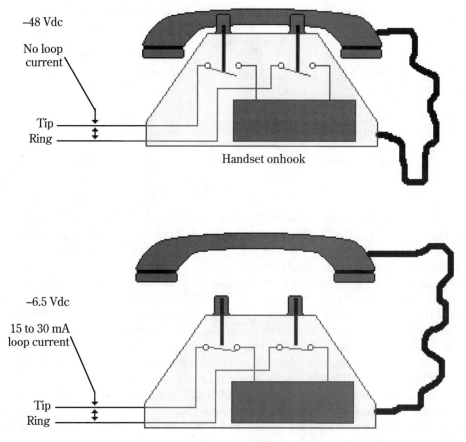

−48 Vdc

No loop
current

Tip

Ring

Handset onhook

−6.5 Vdc

15 to 30 mA
loop current

Tip

Ring

1-10   The mechanics of basic dc signalling.

close and between 15 to 30 mA of dc loop current begins to flow in the local loop. The load presented by your active telephone will cause battery voltage to drop to about −6.5 Vdc as shown in Fig. 1-10. The CO senses this current flow and determines that your telephone is requesting service.

Finally, rotary or pulse dialing (as you saw in the previous section) makes use of loop current to specify a desired number. Rotary dial contacts are made to open and close at a constant rate. When the contacts are open, it causes a brief interruption in loop current—just as if you had hung up the telephone. With the proper timing provided by the dial mechanics, the CO will interpret such interruptions as digits.

Ac signals

There are many audible tones used in local loop signalling. Some tones are used to control the network, while others indicate the network status conditions. Most tones are generated as frequency combinations, which are easier to detect audibly and harder to produce accidentally as false signals. A brief table of status tones is presented in Table 1-1.

Table 1-1. A general listing of telephone status tones.

Signal Name	Frequencies (Hz)	Duration (on/off in seconds)
Dial tone	350 & 440	Continuously on
Busy tone	480 & 620	0.5 on / 0.5 off
Ringback tone	440 & 480	2 on / 4 off
Toll congestion	480 & 620	0.2 on / 0.3 off
Hang up alert	1400 & 2060 & 2450 & 2600	0.1 on/ 0.1 off **Loud tone*
No such number	200 & 400	Continuously switched between 200 Hz and 400 Hz.

The most familiar control tones are those used for DTMF signalling. As you saw in Fig. 1-9, seven distinct frequencies are available from the DTMF keypad, ranging from 697 Hz to 1477 Hz. When one of the 12 keys are pressed, two of these seven tones are combined to represent the desired digit. When a telephone number must be transmitted between various telephone facilities, numbers from 0 to 9 can be represented using only six frequencies (700 Hz, 900 Hz, 1100 Hz, 1300 Hz, 1500 Hz, and 1700 Hz) in pairs, much like DTMF signalling. This is just one of many "interoffice" signalling techniques. Another very common signal (although you might not consider it to be a tone) is the 20-Hz ringing signal generated by the CO.

Status signals keep you informed as to the network's state. The most common status tone is called the *dial tone*. When you first request service from the local CO by picking up your handset, the CO connects a dial tone signal to your line. This tells you that the CO has acknowledged your request for service and is ready to accept the desired telephone number from your dialer. Once the first digit is received, the CO will disconnect its dial tone signal and await subsequent digits. Dial tone is a continuous signal made by combining 350 Hz and 440 Hz tones.

If the CO servicing the desired telephone senses that it is currently in use (drawing loop current), the originating CO will send a "busy" signal back to the calling telephone. A busy signal is a combination of 480-Hz and 620-Hz tones, which is switched on for 0.5 seconds, then switched off for 0.5 seconds. If the CO servicing the desired telephone determines that it is currently idle (not drawing loop current), it will send the appropriate ringing signal to indicate the presence of an incoming call.

The originating CO also sends a ring confirmation tone, called a *ringback tone*, back to the calling telephone. A ringback tone mixes 440-Hz and 480-Hz tones. It is switched on and off to mimic the ringing cadence of the destination CO. For example, if you call a number in the United States, the ringback tone will be switched on for 2 seconds, then off for 4 seconds. If you call a number in another country, your ringback tone will resemble that country's particular ringing cadence.

There is a finite number of telephone lines that interconnect various offices and centers in the PSTN. These are called *trunk lines*. If all the trunk lines are in use, the call cannot go through. When this happens, the originating CO informs the calling telephone by returning a *toll congestion* tone. This is a mix of 480-Hz and 620-Hz tones that is on for 0.2 seconds, and off for 0.3 seconds. It is then necessary to clear the line, and try again later.

A telephone must be onhook when not in use. This ensures that connections and facilities are clear so that other calls can go through. Once a telephone goes offhook, it is given a limited amount of time to begin dialing. If dialing does not begin within that time, the dial tone is shut off and a very loud alarm tone (called a *hang up tone*) is activated. The hang up tone mixes four frequencies: 1400 Hz, 2060 Hz, 2450 Hz, and 2600 Hz. It is switched on and off at 0.1-second intervals. The hang up tone is also activated if the telephone does not hang up after a conversation is completed.

Although a CO is typically able to handle 10,000 telephones, that does not necessarily mean that 10,000 telephones are connected to it. In fact, it is virtually impossible to find a CO that is utilized 100%—many potential numbers are simply not in service. When a called telephone does not exist, an alarm tone is generated. This is a continuous tone that switches back and forth between 200 Hz and 400 Hz in 1 second cycles.

Digital signals

A form of signalling that is rapidly growing in popularity is digital signalling. The term *digital* indicates that the signal represents only two conditions: on or off (true or false). Dc signalling is a fundamental form of digital signalling because loop current is switched on or off depending on whether the handset is onhook or offhook. Digital signalling, on the other hand, uses on/off conditions in organized sequences to represent letters, numbers, or commands.

Instead of interrupting loop current to provide these on/off states, two fixed tones are often used. One tone represents a digital "true," while a secondary tone represents a digital "false" condition. Two dc voltage levels can also be used for digital signalling. The presence of a certain voltage can represent a "true" state, while the absence of that voltage can represent a "false" state. Figure 1-11 shows a simple example of a digital signal. When these binary digits (or bits) are transmitted in the proper time frame, they can carry valuable information quickly and efficiently through the PSTN.

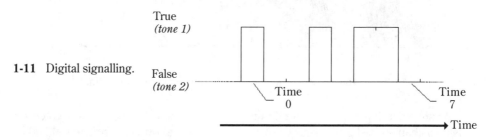

1-11 Digital signalling.

Digital signals have been used for years in interoffice signalling, but only now are they starting to find applications in the local loop for such features as Caller ID. Caller ID allows the called telephone to view the number of the calling telephone. Digital signalling is required to pass this information to a receiving unit at the called telephone.

Transmission characteristics

Ultimately, the purpose of the PSTN is to transmit information from one place to another. For the purposes of this book, you can assume that all this information is made up of analog voice signals. In practice, however, a great deal of computer data from such devices as modems or facsimile machines also travels on the PSTN. This section introduces the electrical characteristics and limitations of a typical telephone voice channel.

Normal human speech is composed of many different frequencies, usually ranging from below 100 Hz to above 6000 Hz. If speech is reproduced with this entire range of frequencies, it will contain all the information of the original speech and be fully intelligible. However, not all the frequencies within this range are needed to produce intelligible speech. Research has shown that the greatest majority of speech information is contained in frequencies between 200 Hz to 4000 Hz. Voice channels on the PSTN are designed to carry frequencies between 0 Hz to 4000 Hz. This is also referred to as a *message channel*, or a VC (voice channel). The range of frequencies that such a channel will pass is known as *bandwidth*. The frequency breakdown of a typical telephone line is shown in Fig. 1-12.

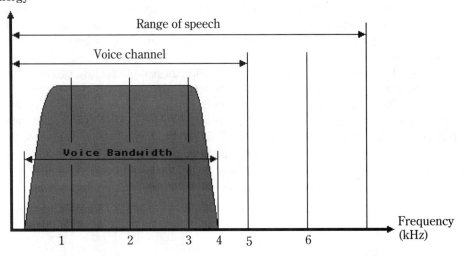

1-12   The typical voice channel configuration of a telephone line.

As shown in Fig. 1-12, the actual range of frequencies allowed to carry voice signals on a channel is slightly narrower than the channel allows for, from about 300 Hz to only 3000 Hz. The voice channel area above 3500 Hz is reserved for special control signals. You might wonder what happens to the voice frequencies from 3000 Hz to 6000 Hz that are not carried on the voice channel. Very simply, these frequencies are never transmitted. Because little real information is included in this range, it can be eliminated without losing speech intelligibility. You can, however, tell some of the information is lost by the often flat or bland tone of the speaker's voice.

Decibels in telephone transmission

There is little work in telephone electronics that does not involve the use of decibels. Although this book is not intended to teach math or engineering, it is helpful if you have a fundamental understanding of decibels and how they are used in the telephone industry. Many of your telephone's specifications are rated in decibels.

In audio electronics—telephones, in this case—it is often interesting to know the amount of power that is gained or lost in a system, or at various points within that system, relative to some reference level. For example, you might like to know the amount of gain or loss provided by a circuit or cable. Such relationships are expressed in decibels.

A decibel is little more than a ratio, such as the power into an amplifier versus the power out of the amplifier. That's all there is to it, except that the ratio is expressed as a logarithm instead of a simple number. There is no real mystery to the logarithm. It is simply a mathematical function (like multiplication, squares, or square roots). Most scientific calculators have a logarithm function built right in.

Consider the amplifier example in more detail. Suppose 1 watt of power went into the amplifier and 100 watts of power was available at its output. By taking a ratio of output power (Po) to input power (Pi), the amplifier has a power ratio of

$$[Po/Pi = 100/1 =]100$$

There are no units because both "watts" units cancel out. However, this is not an expression of decibels. To convert this number to decibels, you can use the relationship shown in Fig. 1-13. By plugging in each power value, that same amplifier would offer a gain of:

$$[10 \times \log_{10}(100/1) =] \ 20 \ \text{decibels (or dB)}.$$

Find your ratio first (Po/Pi), take its log next, then multiply that result by 10. It's that easy.

$$dB = 10[\log_{10} (Po/Pi)]$$

1-13 The basic mathematical equation for the decibel.

where:

---

dB is in decibels
Po is power out in watts
Pi is power in in watts

What is the advantage of expressing a ratio of 100 as 20 dB? Well, there is no real advantage for this particular example, but there are two situations when decibels prove handy. First, they allow very large ratios to be expressed as manageable numbers (a ratio of 100,000 would only be 50 dB). Second, small fractional ratios can be expressed as manageable negative numbers because logarithms of ratios less than 1 have a negative result (a ratio of 0.0001 would be –40 dB). Decibel numbers that are positive represent a "gain," while negative decibel numbers indicate a "loss."

A special ratio is created when an input power of 1 milliwatt (mW) is used as a standard input power. This means that power ratios can now be sized as decibels rated to 1 mW (or dBm). This measurement is particularly handy in the telephone system, because a 1 mW reference level can easily be set at any point in the system. This reference point is then known as the "zero transmission level point" (0TLP) of the system. All other gains and losses can be measured relative to that point. When measurements are made relative to the 0TLP, results would be marked dBm0. Keep in mind that all these measurements are still in decibels even though their labels might be different.

2
Components

A WIDE VARIETY OF ELECTROMECHANICAL AND ELECTRONIC COMPONENTS MAKE
up your telephone—and troubleshooting is much easier if you can identify these components on sight (Fig. 2-1). This chapter introduces you to a cross section of electromechanical and electronic parts, and briefly explains their construction and
operation. This is by no means intended to be a complete course on every possible
electronic part, but it will give you a good idea of what to expect.

Electronic parts

Most of the components that you will encounter in a telephone are electronic. Parts
such as resistors, capacitors, and inductors are generally classified as *passive* components. They are called passive because their only purpose is to dissipate or store
electrical energy. By themselves, these parts serve little practical purpose.

   Components such as diodes, transistors, and all forms of integrated circuits fall
under the much broader category of *active* components. Active components use a circuit's energy to perform a function or a set of functions. The function can be as simple as a diode's rectifier action, or as intricate as a microprocessor's computations,
but all active parts have one thing in common—they do something. This part of the
chapter reviews a selection of both active and passive components.

Resistors

All resistors serve one purpose—they dissipate power. While this might sound waste-
ful, the effects of resistance are actually quite useful and important. Resistors are commonly used to limit or reduce the amount of current flowing in a circuit. Several

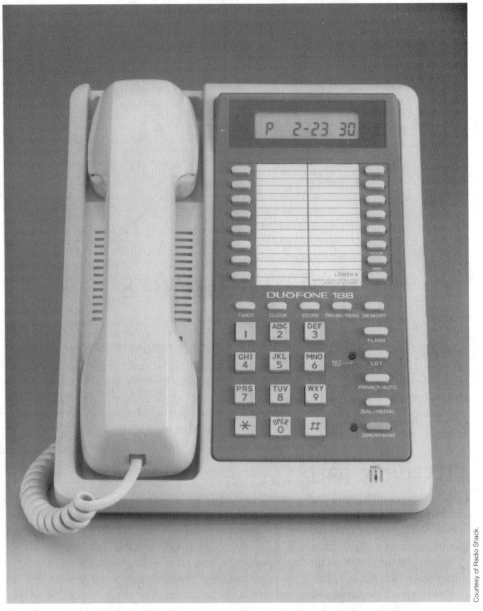

Courtesy of Radio Shack.

2-1 An "intelligent" telephone with enhanced memory dialing features.

resistors can be combined in series to divide voltage into levels necessary for other components, such as transistors or integrated circuits, to function properly. Resistors can also be combined in parallel to divide a source of current into smaller, more useful levels.

Resistors dissipate power by presenting resistance to the flow of current (electrons) passing through them—the unwanted energy is given up as heat. Resistance is measured in ohms and is represented by using the Greek symbol omega (Ω).

Resistors are manufactured by packing carbon filling into a container as shown in Fig. 2-2. Because it is much harder for electrons to pass through carbon than through copper, current flow is limited. The material composition of the resistive filling can be altered to achieve just about any amount of resistance.

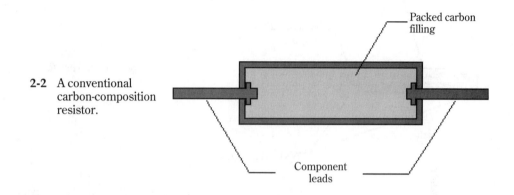

2-2 A conventional carbon-composition resistor.

Packed carbon filling

Component leads

Another common type of resistor is the carbon-film type as shown in Fig. 2-3. Instead of using carbon granules, a glass or ceramic core is layered with a coating of carbon film. The thickness of this coating affects the amount of resistance—thicker coatings yield lower levels of resistance, and vice versa. Metal caps are placed on either end to provide the electrical connections, and the entire assembly is coated in a hard epoxy. Carbon-film resistors are generally more accurate than carbon-composition resistors because a film can be deposited with more precision and control.

With the explosive growth of surface mount technology (SMT), it is helpful to review a surface mount carbon-film resistor as shown in the cross-sectional view of

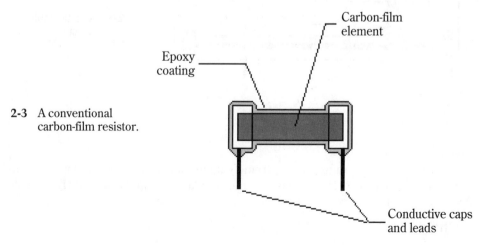

2-3 A conventional carbon-film resistor.

Carbon-film element

Epoxy coating

Conductive caps and leads

Fig. 2-4. An SMT resistor is formed much like a carbon-film resistor—carbon film is deposited onto a nonconductive glass or ceramic substrate placed between two metallic electrodes. The layer's thickness controls the amount of resistance. SMT resistors are incredibly small, ranging from about 1 to 6 millimeters (mm) in length.

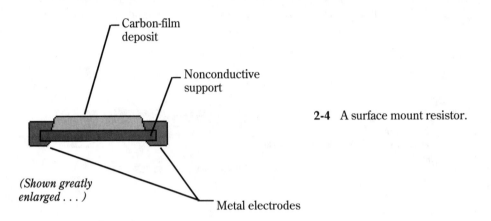

2-4   A surface mount resistor.

Adjustable resistors, known as *potentiometers* or *rheostats*, are usually employed to adjust volume or some other circuit operating parameter. As shown in Fig. 2-5, a typical potentiometer consists of a moveable metal wiper resting on a layer of resistive film. Although the total resistance of the film, end-to-end, remains unchanged, resistance between either end and the wiper blade will vary as the wiper is moved.

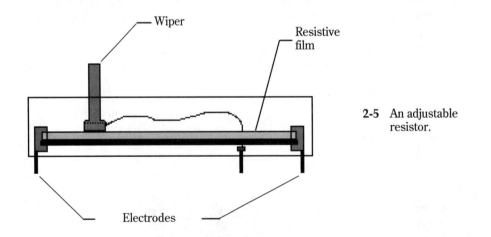

2-5   An adjustable resistor.

Power dissipation ratings are extremely important with resistors. Power is normally measured in watts (W), and is dependent on the amount of current (I) and voltage (V) applied to the resistor as given by the expression:

$[P = I \times V]$.

Resistors are typically manufactured in ⅟₁₆-, ⅛-, ¼-, ½-, 1-, 2-, and 5-watt sizes to handle a wide variety of power conditions. Size is directly related to power dissipation ability, so larger resistors are generally able to handle larger amounts of current than a smaller resistor of the same value.

As long as power dissipation is kept below its rating, a resistor should hold its resistance value and perform indefinitely. However, when a resistor is forced to exceed its power rating, it cannot shed heat fast enough to maintain a stable temperature. Ultimately, the resistor will overheat and burn out as shown in Fig. 2-6. A burned-out resistor always forms an open circuit. A faulty resistor might appear slightly discolored, or it can appear burned and cracked. It really depends on the severity and duration of its overheating. Extreme overheating can burn a printed circuit board. Replace any faulty resistors wherever you find them.

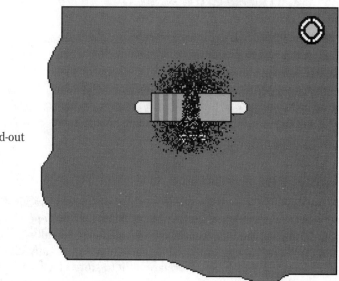

2-6 A burned-out resistor.

Potentiometer failure usually takes the form of intermittent connections between the wiper blade and resistive film. Film wears away as the potentiometer's wiper moves back and forth across it. Over time, enough film might wear away at certain points that the wiper cannot make good contact there. This can result in erratic or intermittent operation. Replace any intermittent potentiometers or rheostats.

Capacitors

Capacitors are simply energy storage devices. They store energy in the form of an electrical charge. The effects of capacitance have important applications in circuits such as filters, tuners, oscillators, and power supplies—just to name a few. Capacitance is measured in farads (F). In actual practice, a farad is a very large amount of capacitance, so most normal capacitors measure in the microfarad (μF) or picofarad (pF) range.

In principle, a capacitor is little more than two conductive plates separated by an insulator called a *dielectric*, as shown in Fig. 2-7. The amount of capacitance provided by this type of assembly depends on the area of each plate, their distance apart, and the dielectric material that separates them. When voltage is applied to a capacitor, electrons flow into it until it is fully charged. At that point, current stops flowing, even though voltage might still be applied—and the voltage across the capacitor will equal its applied voltage.

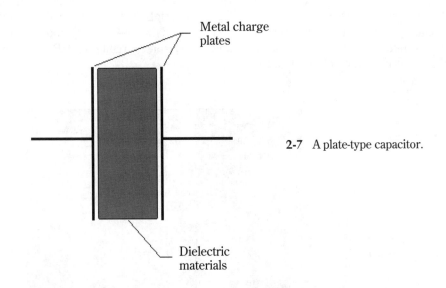

Metal charge
plates

2-7   A plate-type capacitor.

Dielectric
materials

If applied voltage is removed, the capacitor tends to retain the charge of electrons deposited on its plates. Just how long it can do this depends on the specific materials used to construct the capacitor, as well as its overall size. Internal resistance through the dielectric material will eventually bleed off any charge. All you really need to remember is that capacitors are built to store electrical charges.

You should be familiar with two types of capacitors—fixed and electrolytic. Both types are illustrated in Fig. 2-8. Fixed capacitors are nonpolarized devices, which means they can be inserted into a circuit regardless of their lead orientation. Many fixed capacitors are assembled as small wafers or disks. The conductive plates are typically made of aluminum foil. Common dielectrics include paper, mica, and various ceramic materials. The complete assembly is then coated in a hard plastic, epoxy, or ceramic housing to keep out humidity. Larger capacitors are often assembled into large, hermetically-sealed canisters.

Electrolytic capacitors are polarized components—they must be inserted into a circuit in the proper orientation with respect to the applied signal voltage. Electrolytic capacitors are shown in Fig. 2-8. Tantalum electrolytic capacitors are often manufactured in a dipped (teardrop) shape—or as small canisters. Aluminum electrolytic capacitors are commonly used in general-purpose applications where polarized devices are needed. The difference between fixed and electrolytic capacitors is primarily in their materials—the principles and purpose of capacitance remain the same.

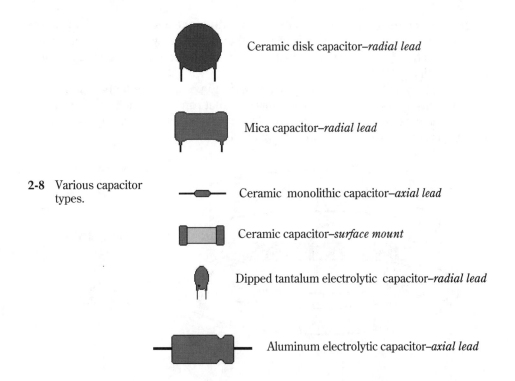

Ceramic disk capacitor–*radial lead*

Mica capacitor–*radial lead*

2-8 Various capacitor
types.

Ceramic monolithic capacitor–*axial lead*

Ceramic capacitor–*surface mount*

Dipped tantalum electrolytic capacitor–*radial lead*

Aluminum electrolytic capacitor–*axial lead*

Capacitors are often designated as *axial* or *radial*. These terms simply refer to the capacitor's particular lead configuration. When both leads emerge from the same end of the capacitor, the device is said to be radial. If the leads emerge from either side, the capacitor is known as axial.

Surface mount capacitors are usually fixed ceramic devices that have a dielectric core and are capped by electrodes at both ends. If an electrolytic capacitor is needed, a surface mount tantalum device is typically used. Although the construction of a surface mount tantalum capacitor differs substantially from a ceramic surface mount capacitor, they both appear very similar to the unaided eye. All polarized capacitors are marked with some type of polarity indicator.

Like resistors, most capacitors tend to be rugged and reliable devices. They only store energy and do not dissipate it, so it is virtually impossible to burn them out. However, capacitors can be damaged or destroyed by exceeding their working voltage (WV) rating, or by reversing the orientation of a polarized device. This can occur if a failure elsewhere in a circuit causes excessive energy to be applied across a capacitor, or if you install a new electrolytic capacitor incorrectly.

Inductors

Inductors, like capacitors, are simply energy storage devices. They store energy in the form of a magnetic charge. Advances in solid-state electronics have rendered inductors virtually obsolete in traditional applications such as resonant (or tuning) circuits or oscillators—but they remain invaluable in telephone applications for impedance matching and voice signal coupling within the telephone itself. Inductors are also indispensable in power supply and other high-energy circuits. Inductance is measured in henries (H)—small inductors are manufactured in the millihenry (mH) or microhenry (µH) range.

A simple inductor, such as the one shown in Fig. 2-9, is little more than a coil of wire. In many cases, a *permeable core* (material that can be magnetized) is included in the assembly to increase the strength of its magnetic field. If a core is omitted, the inductor is generally known as an *air core* inductor.

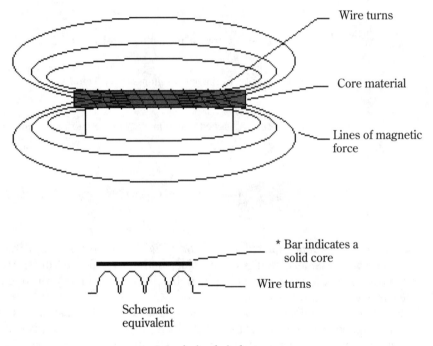

Wire turns

Core material

Lines of magnetic force

* Bar indicates a solid core

Wire turns

Schematic equivalent

2-9   A simple inductor.

When an inductor is energized, it forms a magnetic field. The strength of this field depends on three factors: the number of wire turns in the coil, the amount of current flowing in the coil, and the core material itself (if any). More turns of wire, more energizing current, or a more permeable core material all will result in a stronger magnetic field. The magnetic field generated by inductors is the key to electromechanical ringers used in all classical telephones. Another common application for inductors is the *transformer*.

A transformer is actually a combination of inductors working together. Figure 2-10 illustrates a basic transformer and its three important parts: a primary (or input) winding, a secondary (or output) winding, and a core structure of some type. The operation of a transformer is very straightforward. An ac signal—usually a voice signal in telephone applications—is applied across the telephone's primary windings.

Because the magnitude of the input signal is constantly changing, the magnetic field it generates constantly fluctuates as well. When this fluctuating field intersects

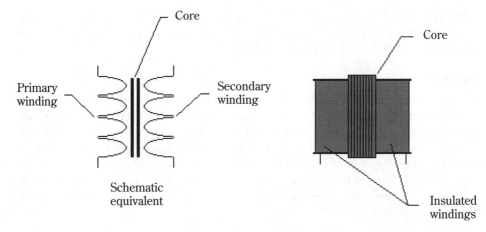

2-10 A basic transformer.

the secondary coil, another ac signal is created, or induced, across it. This principle is known as *magnetic coupling*. Any secondary ac signal will duplicate the original signal. Primary and secondary windings are often wound around the same core structure to provide efficient magnetic coupling from the primary to secondary winding.

Transformers can also be used to mix together more than one signal. A typical telephone application involves the handset transmitter and receiver. Each is a two-wire device, so there are four wires at your handset. The four wires must be converted into a two-wire signal that can be applied over a conventional two-wire telephone line (tip and ring).

A multi-winding transformer, such as the one shown in Fig. 2-11, can be used to interface a four-wire handset circuit to a two-wire line. The associated components and circuitry involved are covered in chapter 6. Notice that each winding is electrically separate from another but every winding is assembled on the same core. Also

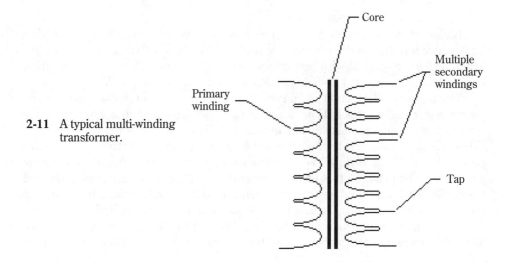

2-11 A typical multi-winding transformer.

notice that a winding can contain one or more access wires (or taps). The transformers found in telephone applications handle very little power. As a result, they typically have high reliability and long operating lives.

Diodes

Diodes are two-terminal semiconductor devices that allow current to flow in one direction, but not in the other. It is this one-way property, known as *rectification*, that makes modern power supplies possible. Because diodes only conduct current in one direction, they are called *polarized* devices. Glass-cased diodes, usually made with silicon, are generally used for low-power small-signal applications. Plastic-cased diodes are typically employed for medium power applications such as power supplies, circuit isolation, and inductive flyback protection. Component views and a schematic symbol are shown in Fig. 2-12.

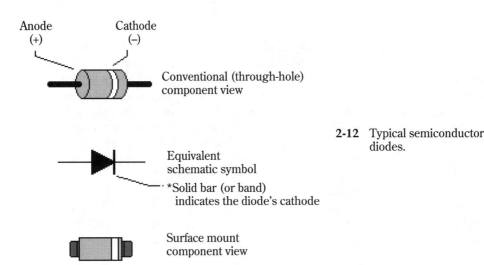

2-12   Typical semiconductor diodes.

A diode has two terminals. The *anode* is its positive terminal, and the *cathode* is its negative terminal. A diode's cathode is always marked with a solid stripe or bar. Most diodes are marked with part numbers beginning with 1N, followed by one to four additional numbers. For example, a diode can be marked: 1N1234.

The internal construction and physical operation of semiconductor devices is beyond the scope of this book. However, it can aid your troubleshooting to understand what circuit conditions cause diodes to conduct. Such a circuit is shown in Fig. 2-13. When a positive voltage is applied across the diode, it conducts current as soon as the potential across it exceeds 0.6 Vdc (for a silicon diode). In this state, the diode is *forward biased*. You can see that the positive portions of the signal are passed through the diode. When a diode is forward biased, it acts as if it were a switch. It cannot limit current flow by itself, so resistance must be added to the circuit to limit current.

If a negative voltage is applied across a diode, it will not conduct current at all. This state, known as *reverse biasing*, is also shown in Fig. 2-13. Notice that the nega-

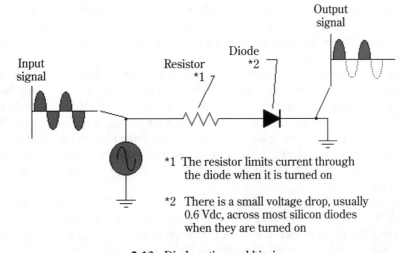

*1 The resistor limits current through the diode when it is turned on

*2 There is a small voltage drop, usually 0.6 Vdc, across most silicon diodes when they are turned on

2-13 Diode action and biasing.

tive portions of the input signal are missing. Ultimately, very large amounts of reverse bias could cause the diode to break down, but as long as voltage remains within safe limits, the diode simply acts as an open switch.

Electronic telephones make extensive use of diodes to rectify the telephone line's dc voltage, as shown in Fig. 2-14. Many electronic telephones use this voltage to power their internal integrated circuits. The diode array illustrated in Fig. 2-14 is called a *bridge rectifier*. Both wires from a telephone line (tip and ring) provide dc voltage from the central office. The rectifier's action ensures that the voltage supplied to the telephone's circuits will be of the proper polarity—even if the tip and ring leads are reversed.

For example, if the voltage at tip is more positive than the voltage at ring, diode D1 will conduct the voltage at tip to the rectifier's output. If the line polarity were reversed,

Voltage from the telephone line is positive regardless of the polarity across tip and ring

2-14 A diode bridge rectifier circuit.

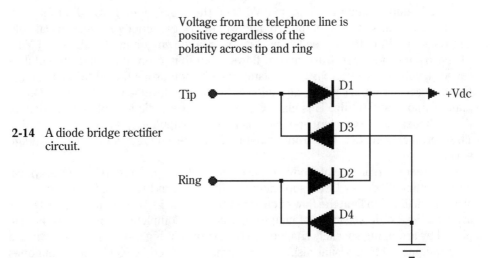

the voltage at ring would be more positive than the voltage at tip. As a result, diode D1 would be off, and diode D2 would turn on to conduct voltage to the rectifier's output. Diodes D3 and D4 are included to provide a common reference point (ground) in relation to the rectifier's output.

Rectifier diodes are meant to be operated under forward bias conditions—they will not conduct in the reverse direction unless they are damaged. Some diodes, however, are meant to operate in the reverse direction. These are known as *zener* diodes. Figure 2-15 shows a simple circuit fragment using a typical zener diode. When the applied voltage is below the zener's *breakdown voltage* (the point when a zener diode begins to conduct current in the reverse direction), the output voltage across the diode equals the applied voltage, and the zener acts just like any other reverse-biased diode.

Zener diodes are meant to conduct at reasonably low voltages (for example: 5.6 Vdc, 9.3 Vdc, 12.6 Vdc, 15.6 Vdc, etc.), so as applied voltage exceeds this breakdown voltage level, the zener begins to conduct. This creates a path of current through the zener, and generates a voltage drop across the current-limiting resistor that maintains a constant voltage across the zener.

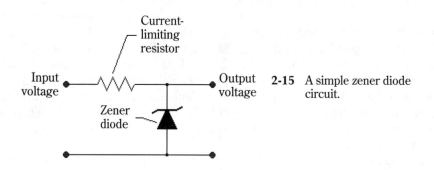

2-15   A simple zener diode circuit.

For example, if input voltage is 2 Vdc and the zener diode is rated for 5.6 Vdc, the diode remains off, and output voltage will follow the input voltage. If input voltage rises to 8 Vdc, this exceeds the zener's breakdown voltage by about 2.4 Vdc. This activates the zener. Current now flows through it. This additional current flow causes a voltage drop across the resistor, which then drops the additional voltage (in this case, about 2.4 Vdc)—and allows the voltage level across the zener to remain clamped at 5.6 Vdc. This clamping action is the basis for voltage regulation. Zener diodes have important applications in power supplies and telephone circuits. They protect sensitive ICs from variations in line voltage, and severe voltage surges.

Another popular type of diode produces light when it is forward biased. A typical light-emitting diode (or LED) application is illustrated in Fig. 2-16. Notice that a resistor is included to limit the flow of current through the LED when it is on. LEDs are available in a wide selection of shapes and colors. Complete numbers can be displayed by arranging specially-shaped LEDs as shown in Fig. 2-17. This is known as a seven-segment LED. Visual displays are commonly used in electronic telephones

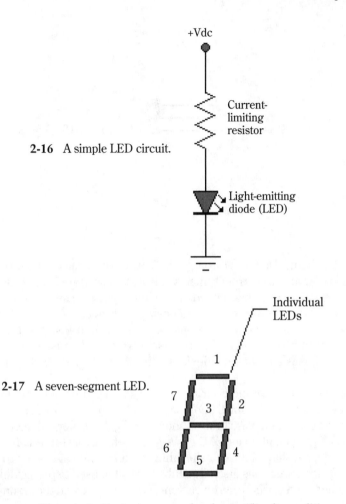

2-16 A simple LED circuit.

2-17 A seven-segment LED.

and answering machines to indicate power, onhook or offhook conditions, the number of calls waiting, and more.

Surge suppression

Lightning strikes, power line induction, and even rotary dialing pulses can at times produce voltage spikes in excess of 1000 volts. This much voltage will destroy transistors and ICs almost instantly, so any such surges must be removed from the telephone line before they reach your telephone.

In the last section, you learned that zener diodes can be used to clamp dc voltages to certain, fixed levels. Not only is this useful for providing necessary voltages to ICs and other circuits, but high-voltage zener diodes can be used to clamp (or short-circuit) any high-voltage noise or transient signals that might occur across a telephone line.

Two other circuit protection devices that you should be familiar with are the fuse and the gas discharge tube. Fuses limit the amount of current entering a circuit. A filament of fine metal wire is strung across two metal ends capping a clear glass enclosure, as shown in Fig. 2-18. The fuse is inserted in series with the circuit, so the current

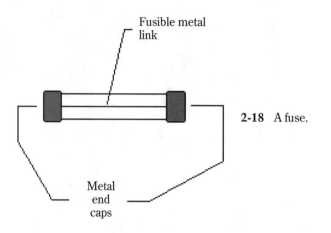

2-18 A fuse.

drawn by the circuit also flows through the fuse. Current causes a temperature rise in the metal link. As long as current remains below the fuse's rated limits, the link remains intact. However, if current exceeds the fuse's rated value, heat buildup causes the link to open—this cuts off power to the circuit being protected.

A fuse responds too slowly to protect a circuit from transients, but if a transient should damage a component and cause a short circuit, the fuse will open to prevent a fire hazard. Fuses are one-time devices—they must be replaced if their metal link opens.

For large or long-duration surges, gas discharge tubes can be used. Gas discharge tubes are much like small fluorescent light bulbs—electrodes are separated by inert gas in an evacuated glass tube as shown in Fig. 2-19. Normal telephone line voltages are not high enough to ionize the gas, so it is effectively turned off. When a surge does occur, high voltage ionizes gas within the tube and causes a current flow that short-circuits the surge. Gas discharge tubes are often used to protect telephone lines from lightning strikes, where surge currents can easily exceed thousands of amperes.

SCRs

Silicon-controlled rectifiers (SCRs) are somewhat more sophisticated than ordinary rectifier diodes. As you can see in Fig. 2-20, a SCR uses a third wire (called a gate) to control its actions.

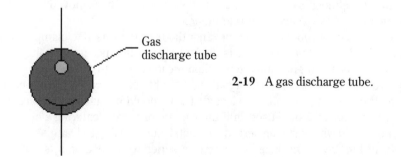

2-19 A gas discharge tube.

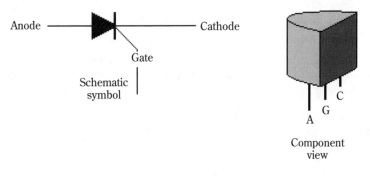

Component
view

2-20   A silicon-controlled rectifier (SCR).

Until a control voltage is applied to the gate, the SCR is turned off—that is, no current will flow in either direction, regardless of how the SCR is biased (anode to cathode). When gate voltage is applied, the diode will turn on and behave just as any ordinary rectifier diode would. Once a SCR is turned on and current is flowing, the gate voltage can be removed. The diode will continue to conduct until its bias is removed. This shuts down the SCR again until gate voltage is reapplied. SCRs are particularly useful in electronic telephones for line hold functions—especially in multiline electronic telephones.

Transistors

A transistor is a three-terminal semiconductor device whose output signal is directly controlled by its input signal. This makes transistors particularly well-suited for signal amplification and switching tasks—both are necessary in electronic telephones.

There are three major families of transistors, as shown in Figs. 2-21, 2-22, and 2-23. Bipolar transistors are common, inexpensive, general-purpose amplifiers and switches. Many electronic telephone circuits that use discrete (individual) transistors use bipolar devices. For the purposes of this book, bipolar transistors are shown

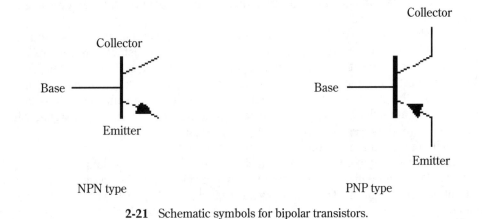

NPN type   PNP type

2-21   Schematic symbols for bipolar transistors.

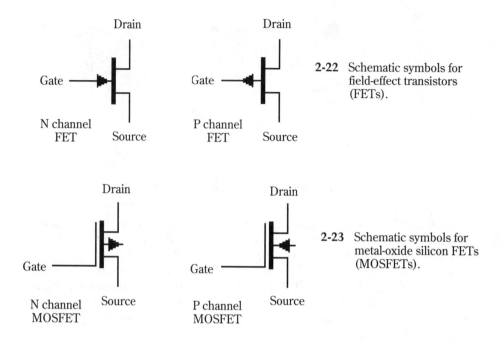

2-22 Schematic symbols for field-effect transistors (FETs).

2-23 Schematic symbols for metal-oxide silicon FETs (MOSFETs).

in most circuit examples. Discrete field-effect transistors (FETs) and metal-oxide semiconductor FETs (or MOSFETs) are rarely used in telephones as individual components, although they are used extensively in integrated circuits. There is not enough room in this book to discuss the characteristics and operation of each transistor family, but it is important for you to know that they are all transistors—and be able to identify them on sight.

Transistor part number markings are as varied as transistor families. Standard transistor markings begin with the prefix 2N, followed by up to four other digits. With the vast proliferation of transistors, however, individual manufacturers can prefix their part number with other markings such as MPS, SD, VN, J, TIC, MJ, MJE, TIP, or MPQ. You will need to refer to a manufacturer's data book for electrical and performance specifications.

A basic transistor amplifier is shown in Fig. 2-24. The amplifying transistor is a bipolar NPN-type device. Resistors are used to bias the transistor. This sets up the idle voltage and current conditions that a transistor needs to function as an amplifier. The capacitors placed at the input and output are used to block any dc current that might upset the transistor's bias conditions—only ac signals are allowed to pass into or out of the amplifier circuit. When configured in this fashion, an output signal will be a larger, more powerful version of the input signal.

Switching circuits are somewhat simpler because they do not need the same delicate biasing as amplifiers. A simple transistor switching circuit is illustrated in Fig. 2-25. Unlike an amplifier whose output varies in proportion to its input, a switching circuit is either on or off. The digital signals that activate the switch are usually provided by logic gates. A logic 0 leaves the transistor off, while a logic 1 turns the tran-

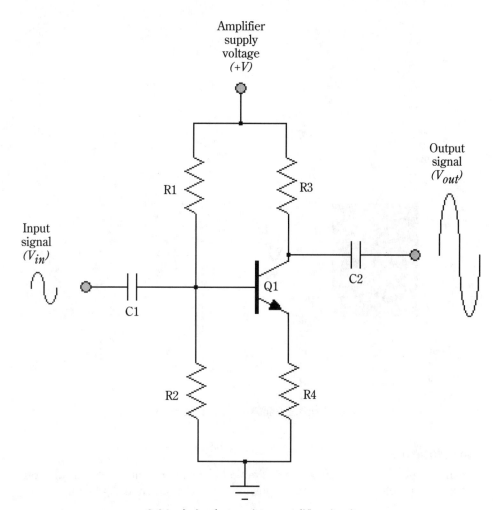

2-24 A simple transistor amplifier circuit.

sistor on. One or more resistors can be added to the base and collector circuits for current-limiting purposes.

Transistors of all types are available in a variety of case styles. Their style depends on the amount of power that the device must dissipate. Figure 2-26 illustrates a selection of three conventional case styles. Low-power devices are often packaged in small, plastic TO–92 cases. Medium-power transistors use the larger TO–220 cases with a metal mounting flange. The flange provides mechanical strength, as well as a secure path for an external heat sink. The TO–3 case is used for high-power transistors. Two mounting holes are provided to bolt the device to a chassis or external heat sink. As a general rule, case size is proportional to the power capacity of the transistor. Transistors are also manufactured in surface mount cases. Two typical surface mount transistor (SOT) case styles are shown in Fig. 2-27.

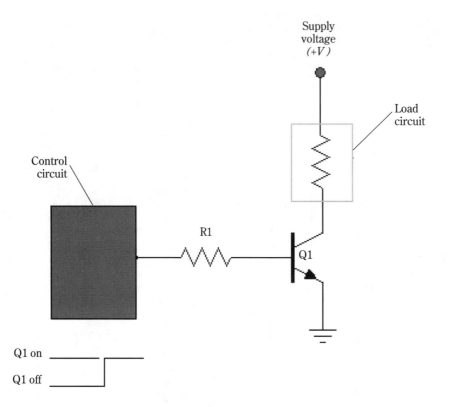

Supply
voltage
(+V)

Load
circuit

Control
circuit

R1

Q1

Q1 on

Q1 off

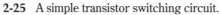

2-25 A simple transistor switching circuit.

Optoisolators

In many telephone applications, it is desirable to isolate one part of the circuit from another. This is often done to protect a low-power circuit (such as digital logic) from

TO-92 case

TO-220 case

TO-3 case

2-26 Three conventional transistor case styles.

E   B

C

* SOT= small outline transistor

E B C

SOT-23 style   SOT-89 style

2-27 Several surface mount transistor case styles.

potential damage by a high-power circuit (perhaps an electromechanical ringer). Isolation can be accomplished optically, using light to carry a signal, without any electrical connection between circuits. A typical optical isolator (or *optoisolator*) is illustrated in Fig. 2-28.

There are two sections to an optoisolator: a transmitter and a receiver. A transmitter is essentially an LED—either visible or infrared (IR). The receiver is a phototransistor that is particularly sensitive to the wavelengths of light generated by the transmitter. When a signal is applied to the transmitter, the LED will turn on and off in proportion to that signal. In many cases, the optoisolator is carrying digital signals, so the LED is either fully on or fully off. Light from the LED stimulates the phototransistor's base. This turns the receiver on in direct proportion to the amount of transmitted light, so the original signal is reproduced at the optoisolator's output.

Optoisolators are commonly used to isolate the high-voltage ring signal on a telephone line from the remainder of the telephone circuitry. They can also be used to create an optical hook switch as shown in Fig. 2-29. A slotted optoisolator is used for this application, and the LED portion is kept on continuously. When the handset is lifted offhook, spring tension pulls the hook switch lever up and out of the slot. Light travels across the gap, activates the phototransistor and creates a logic output for other telephone circuits to interpret. When the handset is replaced on its cradle, its weight pushes the hook switch lever back into the optical slot. This cuts off the light path between transmitter and receiver, resulting in an opposing logic output. An optical hook switch is immune to the mechanical problems associated with ordinary switches.

Transmitter   Receiver

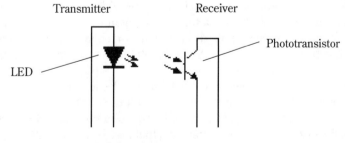

LED   Phototransistor

2-28 A typical optoisolator.

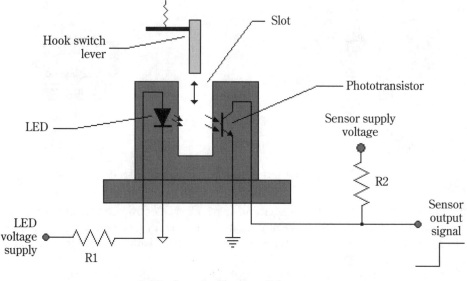

2-29   An optical hook switch sensor.

Liquid crystal displays

Many of the more intelligent telephones available today use multi-segment liquid crystal displays (LCDs) for displaying such things as dialed number, line status, date, time, and etc. Their low power consumption, long working life, and easy-to-form characters have made LCDs the display choice for most electronic telephone applications. Even answering machines and cellular telephones make use of LCDs.

The general construction of a simple, twisted nematic liquid crystal display is shown in Fig. 2-30. The heart of this display is a layer of liquid crystal material sandwiched between two thin glass plates. In its normal state, liquid crystal molecules are twisted, so light entering the front polarizing filter is rotated 90 degrees. This allows light to pass through the rear polarizer, off a mirrored reflector (not shown in Fig. 2-30), and back up through the liquid crystal where it is twisted again and passed through the front polarizer. In this condition, the display appears clear or transparent.

Conductive electrodes are deposited onto the front and rear glass plates. In Fig. 2-30, electrodes are shown as seven-segment digits, but any image could just as easily be used—as long as the same image is provided by each electrode. By giving electrodes a variety of shapes, a full complement of characters and letters can be formed. When a voltage is applied across two corresponding electrodes, an electric field is established through the liquid crystal material—much the same principle as with an ordinary capacitor.

An electric field causes the liquid crystal molecules just under the activated electrodes to straighten out. After light passes through the front polarizer, it is no longer twisted by the liquid crystal, so it is absorbed by the rear polarizer. Because light passing through the crystal is no longer reflected, that area is now visually opaque. Liquid crystal visibility depends on the presence of ambient light that is reflected back to

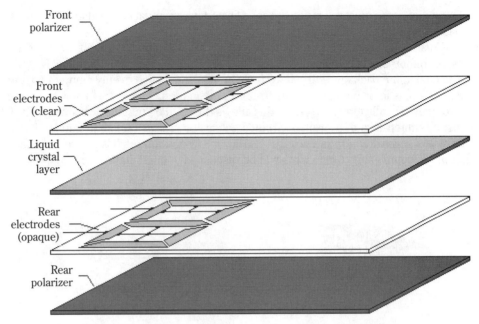

Front polarizer

Front electrodes (clear)

Liquid crystal layer

Rear electrodes (opaque)

Rear polarizer

2-30 A typical LCD assembly using the twisted nematic effect.

form the image. As a result, LCDs cannot be seen well in darkness or low-light conditions. This limitation can often be overcome by backlighting the LCD assembly. When electrodes are activated, backlight is blocked to form an opaque image.

Integrated circuits

Integrated circuits (ICs) are by far the most diverse and powerful group of electronic components that you will ever deal with. They have become the fundamental building blocks of modern electronics. They are used for amplifiers, memories, microprocessors, digital logic, oscillators, regulators, speech circuits, and a myriad of other complex functions. You are bound to find at least one IC in just about every piece of commercial electronics equipment available today.

As a general rule, ICs can be classified as *analog* or *digital*. Analog devices are designed to operate with continuously varying signals such as tones or speech. Digital ICs are primarily used to process binary signals. Memories and microprocessors are two common examples of this. However, the classical distinctions of analog and digital are slowly fading away in light of new, mixed-signal ICs—devices that combine specific analog and digital functions right on the same IC. Mixed signal devices such as this are sometimes called application-specific integrated circuits (ASICs). They are designed and manufactured specifically for use in a particular product or system.

Due to the complexity of ICs, they are normally considered to be *black box* devices—signals are applied and signals are generated, but little consideration is given to what actually occurs inside. From a troubleshooting and repair standpoint, this is usually enough, as long as the IC's purpose and signals are understood. To determine an

IC's functions, you must reference the specific part in a manufacturer's data book, or estimate its functions by evaluating it in a schematic diagram.

ICs are manufactured in a startling variety of package styles. Just a few of the more common styles are illustrated in Fig. 2-31. Probably the most familiar IC style is the dual-inline package (DIP). Two rows of pins are spaced evenly along opposing sides of a plastic or ceramic enclosure. A single-inline package (SIP) has only one row of pins (usually ten or twelve)—this allows the IC to be mounted vertically. They consume much less of a printed circuit's space (or real estate) than similar DIPs. There are several common package styles for surface mount ICs as well. Surface mount PC boards and components will be discussed in detail in the next chapter.

Dual inline
package (DIP)

Dual inline package
surface mount (SMT DIP)

Plastic leaded
chip carrier (PLCC)

Single inline
package (SIP)

2-31   Typical integrated circuit package styles.

Electromechanical parts

Another broad category of parts are classified as *electromechanical*. An electromechanical part can use either mechanical force to initiate an electrical response, or an electrical signal to generate a mechanical force. Keyboards, ringers, transmitters (microphones), and receivers are just a few typical electromechanical components this chapter covers in detail.

Ringers

Ringers are signalling devices that alert you to the presence of a waiting call. In the early days of telephones, it was desirable (from the telephone company's standpoint) that the called party should answer their telephone as soon as possible. This freed the local, toll, and long-distance facilities used to connect the call—and helped to generate income for the telephone company. As a result, original ringers were designed to produce an urgent sound that could be heard at reasonably great distances. Thomas A. Watson patented the first electromechanical ringer for telephone applications in 1878. It used a solenoid coil to drive a metal clapper that strikes a metal gong. Most older telephones still use some version of this design.

Figure 2-32 illustrates a single-gong electromechanical ringer. An ac ringing signal is applied from the CO. The ringing signal is usually 90 Vac at a frequency between 16 Hz to 60 Hz (20 Hz is standard in the United States). This ac signal sets up a fluctuating magnetic field in the solenoid coil, which alternately pushes and pulls the clapper. As the clapper flies forward, it strikes the gong. When the polarity of the ringing signal reverses, it reverses the solenoid's magnetic field and pulls the clapper away from the gong. Notice the capacitor in series with the solenoid. This is used to prevent the solenoid from passing dc—only the ac ringing signal is allowed to pass through. The capacitor and solenoid coil are chosen to give a high impedance to voice frequencies.

The double-gong electromechanical ringer in Fig. 2-33 works in a similar fashion. Instead of one gong, there are two, so the clapper is striking either gong as it is alternately attracted and repelled by the solenoid. Two solenoids can be used in a double-gong ringer.

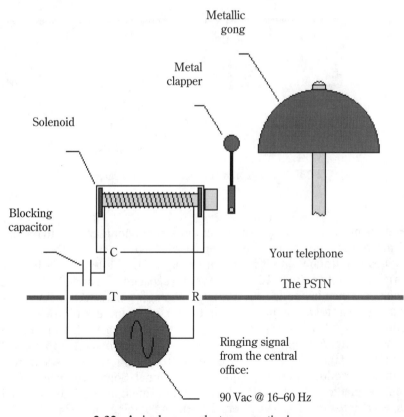

2-32 A single-gong electromagnetic ringer.

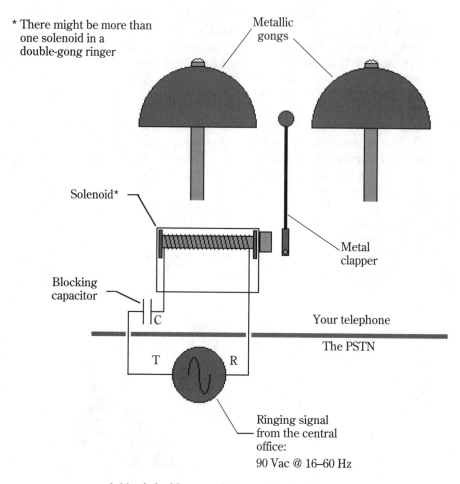

* There might be more than
  one solenoid in a
  double-gong ringer

Metallic
gongs

Solenoid*

Blocking
capacitor

C

Metal
clapper

Your telephone

The PSTN

T   R

Ringing signal
from the central
office:
90 Vac @ 16–60 Hz

2-33   A double-gong electromechanical ringer.

With the proliferation of integrated circuits, it did not take long to devise an IC that could replace the heavy, bulky, and expensive electromechanical ringers. A simplified application circuit is shown in Fig. 2-34. The telephone line (tip and ring) is connected to a ringer IC. A resistor and capacitor are added to limit current and block dc. Once inside the IC, a ringing signal is converted to dc and used to power an oscillator. Oscillator output is then used to drive a piezoelectric transducer that vibrates to produce the desired sound, often more pleasant and soothing than a metal gong. A simple adjustable resistor can then be added to control ringing volume. Today, there are many different tone ringers to select from—each is designed to produce its own unique ringing signal. Some tone ringer functions are even manufactured on the same IC with other functions such as tone/pulse dialers, speech networks, etc. The internal workings of telephone ICs will be covered later.

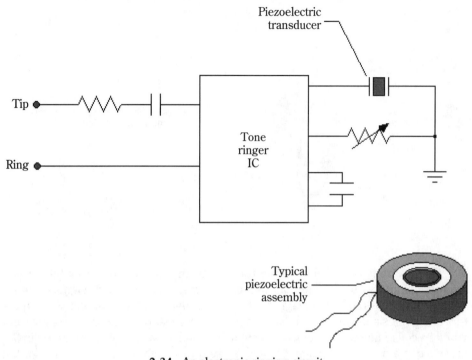

2-34 An electronic ringing circuit.

Transmitters

A transmitter is used to convert the acoustical energy of your speech into a varying electrical signal that can be transmitted through the PSTN to a called telephone. There are three general types of transmitters in use today: carbon microphones, electrodynamic microphones, and electret microphones.

Carbon microphones have changed very little since Thomas Edison first invented them more than 100 years ago. The materials and construction techniques have been improved with time, but their basic principles and construction remain as illustrated in Fig. 2-35. Carbon microphones are still used in older classical telephones.

They are made by filling a two-piece metal container with carbon granules and sealing it together with a flexible, nonconductive coupling. In its simplest form, this is a carbon composition resistor that offers some amount of resistance while it is at rest. The rear portion of this resistive container is a fixed part of the assembly, while the front portion is attached to a rigid metal diaphragm. Both halves of the resistive container are connected to wire leads.

When the acoustical energy in a voice (or other sound) strikes the metal diaphragm, it causes the carbon capsule to vibrate—expand and contract—so its resistance changes in step with the voice signal. When voltage from the telephone line is applied across the transmitter, variations in the transmitter's resistance cause corresponding variations in loop current. In this way, speech signals are introduced to the PSTN.

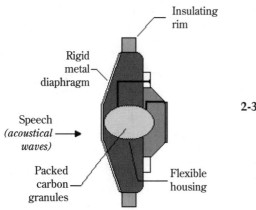

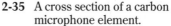

2-35 A cross section of a carbon microphone element.

The electrodynamic microphone represents an improvement over the carbon microphone in its construction, as well as its sensitivity to speech signals. Figure 2-36 shows a cross-sectional diagram of an electrodynamic microphone. Speech enters and strikes a flexible metal diaphragm. This causes the diaphragm to move and vibrate a coil of fine wire wrapped around it. The coil, or voice coil, is mounted in the presence of a permanent magnetic field, so as the coil vibrates, it cuts through the magnetic field to induce a coil current that is proportional to the speech signal itself. This induced speech signal can then be filtered and amplified by the telephone's speech circuits before being introduced to the PSTN. Because of the relatively large

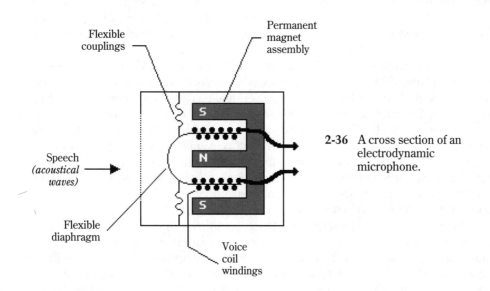

2-36 A cross section of an electrodynamic microphone.

amount of amplification required with an electrodynamic microphone, their use is limited to electronic telephones.

Another type of microphone that is growing in popularity is the electret microphone. An electret microphone, as shown in Fig. 2-37, uses a small piece of dielectric (insulating) material that is metalized on one side. This electret material is then fixed between two metal plates (or rings) to form a sort of capacitor. When speech signals strike the electret material, the vibrations cause voltage to vary at its output terminals. This extremely small output signal must be amplified by the circuits of an electronic telephone. The action of an electret microphone introduces virtually no distortion into the signal. A transistor circuit is used to match the electret microphone's output impedance to that of the amplifier's input.

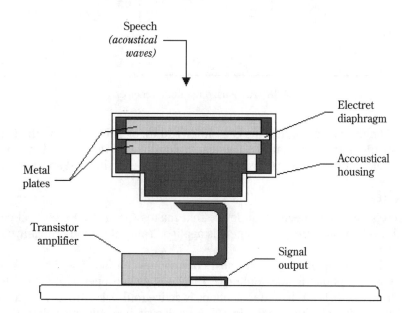

2-37 A cross section of an electret microphone.

Receivers

A telephone receiver element is used to convert electrical speech signals on the PSTN back into acoustical energy (or sound vibrations) that you can hear. Receivers have remained virtually unchanged in design since the first telephone. As shown in Fig. 2-38, a receiver is an electromagnetic device. A permanent magnet assembly is wrapped with a voice coil carrying speech current from the caller. This assembly is then mounted in close proximity to a flexible metal diaphragm. A permanent magnet is used to supply *magnetic bias*—some amount of magnetism in the absence of any speech signals. As speech signal polarity fluctuates in the coil, it affects the magnetic

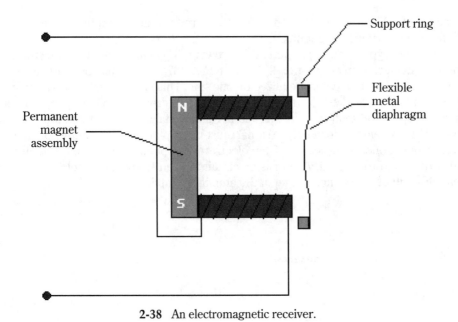

Support ring

Flexible
metal
diaphragm

Permanent
magnet
assembly

N

S

2-38 An electromagnetic receiver.

field holding the diaphragm. This can either add to, or subtract from, the field produced by the permanent magnet. The diaphragm then vibrates in and out to re-create the caller's voice. This is the same basic principle used in most speaker designs.

Switches

A switch is an electromechanical device that makes (or breaks) electrical contacts when a mechanical force (your finger) is applied. Your telephone probably contains a variety of switches.

In telephone applications, the most common switch found is the hook switch. When your telephone is idle (onhook), the hook switch opens to isolate the telephone from the PSTN. When the telephone is active (offhook), its hook switch closes and connects the telephone to the PSTN. A hook switch is little more than a series of spring-loaded electrical contacts activated with the weight of a handset.

Switches can often be broken down into three categories—mechanical, membrane, and contact. Each of these three switch types offers a variety of advantages and disadvantages.

Mechanical switches (metal-on-metal devices) have been around for many years. A simple mechanical slide switch is illustrated in Fig. 2-39. A set of metal terminals, usually plated with tin, are capped with wide metal contacts plated with silver or gold. These terminals are arranged in a row along the switch body. A silver- or gold-plated metal wiper attached to a plastic knob is used to selectively short the desired set of contacts together. By sliding the wiper back and forth, the wiper can be moved to short the center terminal to either the left or right terminal. Mechanical switches are subject to wiper (and contact) wear and corrosion.

Membrane switches are often used when push-button or momentary contact

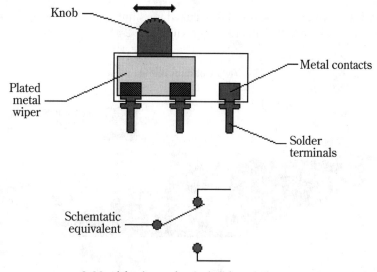

2-39 A basic mechanical slide switch.

functions are needed. A simple membrane switch assembly is shown in Fig. 2-40. A key cap is made with a molded plunger that depresses the switch's rubber cover. When pressure is applied, the rubber cover pushes a metal diaphragm into a conductive block. When pressure is released, the diaphragm returns to normal. The actual travel involved with this type of contact is only several thousandths of an inch. This makes the electrical switch connection. Membrane switches are sometimes used in telephone dial pads. Unfortunately, membrane switches are subject to problems from contact corrosion and diaphragm breakage.

Contact switches have become the preferred replacement to membrane switches. Contact switches are used to create arrays of switches—such as dial pad

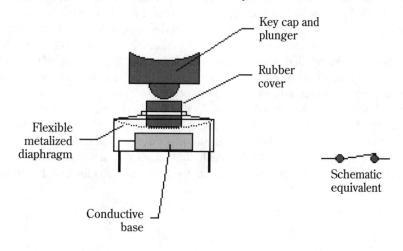

2-40 A membrane switch.

assemblies, so you will never find them used for single push-button applications. The principle behind contact switches is shown in Fig. 2-41.

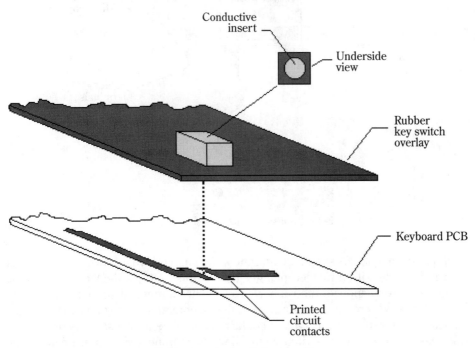

2-41   A contact switch assembly.

Sets of contacts are etched onto a printed circuit board—just as they are on any other kind of PC board assembly. A specially-made rubber overlay, usually in the form of a dial pad, is placed over the PC contacts. Each key in the overlay is aligned over its corresponding contact set. A conductive insert (often plastic impregnated with carbon, or a conductive elastomer) is fitted inside the recess of each rubber key. With no pressure applied to a key, the conductive insert is held off of its printed contacts. When a key is pressed, the rubber key compresses, forcing the conductive insert to bridge its printed contacts. This completes the switch. When pressure is released, the rubber key returns to its normal position, and contact is broken.

Contact switches are easy and inexpensive to manufacture, but you can encounter problems if dirt and dust accumulate on the printed contacts. When contacts become dirty, clean them by wiping each contact set with a cotton swab moistened in a quality electrical contact cleaner. It is a good idea to also clean each conductive insert in the rubber overlay.

3
Service guidelines

ELECTRONIC TROUBLESHOOTING IS A STRANGE PURSUIT—AN ACTIVITY THAT falls somewhere between art and science. Its success depends not only on the right documentation and test equipment, but on intuition and a thorough, careful troubleshooting approach. This chapter shows you how to evaluate and track down telephone problems and locate technical data. It also includes a series of service guidelines to simplify your work.

The troubleshooting process

Regardless of the complexity of your particular telephone (Fig. 3-1) or answering machine, a sound, dependable troubleshooting procedure can be broken down into four basic steps as illustrated in Fig. 3-2:

1. Define your symptoms.
2. Identify and isolate the potential source (or location) of your problem.
3. Repair or replace the suspected component or assembly.
4. Retest the unit thoroughly to be sure you have solved the problem.

If you have not solved the problem, start again from step number 1. This is a universal procedure that you can apply to any sort of troubleshooting—not just telephone equipment.

Define your symptoms

Sooner or later, your telephone or answering machine is going to break down. The cause can be as simple as a loose connector, or as complicated as an IC failure. Before you open your toolbox, you should have a firm understanding of all the symptoms. It's

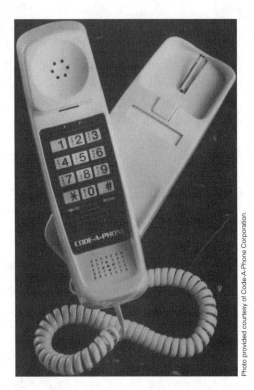

Photo provided courtesy of Code-A-Phone Corporation.

3-1 A simple, hand-held telephone with redial.

not enough to simply say "It's busted!". Think about the symptoms carefully. Ask yourself honestly just what is (or is not) happening. Consider when the problem is occurring. Is the telephone dialing properly? Does the handset have sidetone? Does this problem occur only when the telephone is tapped or moved? If you recognize and understand your symptoms, it is much easier to trace a problem to the appropriate assembly or component.

Take the time to write down as many symptoms as you can—whatever you see or hear (or whatever might be absent). Notetaking might seem tedious now, but once you have begun your repair, a written record of symptoms and circumstances helps to keep you focused on the task at hand. It also helps to jog your memory if you must explain the symptoms to someone else at a later date.

Identify and isolate

Before you try to isolate a problem within a piece of telephone equipment, you must first be sure that the equipment itself is causing the problem. In many circumstances, this is fairly obvious, but there are situations that appear ambiguous (i.e., the equipment never rings, no dial tone is heard, etc.). Always remember that a telephone or answering machine is itself only a small portion of a vast system—the Public Switched Telephone Network (or PSTN). It is possible that a problem in your central office—or somewhere in the miles of interconnecting cables between your central office and your telephone—can be causing the trouble.

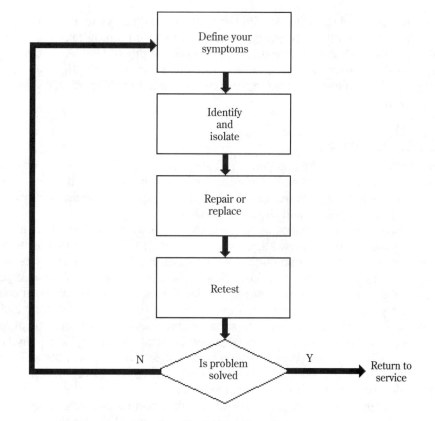

3-2 Troubleshooting flowchart.

This is an easy application of the universal troubleshooting procedure shown in Fig. 3-2. Once you have carefully identified symptoms, isolate your telephone by unplugging it from its wall jack. You can replace the equipment by testing it in another wall jack connected to a known good telephone or answering machine. A friend or neighbor might let you test your equipment in their home.

If your equipment exhibits the same symptoms on a good telephone line, there is an excellent chance that the trouble is in your telephone equipment. You can then proceed with specific troubleshooting procedures for your particular type of equipment. If, however, the symptoms disappear when used on another telephone line, you should suspect a problem in your original telephone line. Refer to chapter 5 for wiring installation and repair procedures. If you cannot locate a wiring problem in your home, contact your local telephone company for service.

Another test is to try known good telephone equipment on your current telephone line. If another telephone or answering machine works properly on your original line, it verifies that your home and outside telephone wiring is working properly. If known good equipment does not work on your line, it is possible that a wiring problem is at the root of your trouble.

When you are confident that your telephone or answering machine is at fault, you can begin to identify the possible problem areas. Start your search at the sub-section level. The troubleshooting procedures described throughout this book will guide you through the major sections of several equipment families, and help you decide which subsection might be at fault. When you have identified a potential problem area, you can begin the repair process—and hopefully track the fault to a component level.

Repair or replace

Now that you have an understanding of what is wrong and where to look, you can begin the actual repair process. Telephones are primarily electronic devices, so most procedures will require the exchange of electronic or electromechanical parts. All procedures are important and should be followed carefully.

Parts are usually classified as *components* or *subassemblies*. A component is the smallest possible individual part that you can work with. Components serve many different purposes in a telephone or answering machine. Resistors, capacitors, transformers, motors, and integrated circuits are just a few types of component parts. Components contain no serviceable parts within themselves—a defective component must be replaced. A subassembly, on the other hand, is composed of many individual components. Unlike components, subassemblies serve a single specific purpose in a telephone. They can usually be repaired by locating and replacing any faulty components. It is always acceptable to replace a defective subassembly with a new one, but complete subassemblies can be very expensive (compared with the cost of a new telephone), and are more difficult to obtain.

Many of the mail-order companies listed in the appendices offer general-purpose electronic components and equipment. Most companies will send their complete catalog(s) or product listing(s) at your request. Keep in mind, however, that consumer electronics products—like telephones and answering machines—make increasing use of specialized integrated circuits and physical assemblies.

For specialized parts, you will often have to deal directly with the manufacturer. Going to a manufacturer is always somewhat of a calculated risk—they might choose to do business only with their affiliated service centers, or refuse to sell parts to consumers outright. If you do find a manufacturer willing to sell you parts, you must know the exact code or part number used by that manufacturer. This information is often available from the manufacturer's technical data—if you have it. Keep in mind that most manufacturers are ill-equipped to deal directly with individual consumers, so be patient and be prepared to make several calls before you find what you need.

During a repair, you might reach a roadblock that requires you to leave your equipment for a day or longer. Make it a point to reassemble your telephone or answering machine as much as possible before leaving it. Gather any loose parts in plastic bags and seal them shut. If you are working with electronic circuitry, make sure to use good-quality anti-static boxes or bags for storage. Reassembly will prevent a playful pet, curious child, or well-meaning spouse from accidentally misplacing or discarding parts while the telephone equipment is on your workbench. This is twice as important if your work space is in a well-traveled or family area. It will also help you remember how to assemble the equipment later on.

Retest

When a repair is complete, the telephone or answering machine must be reassembled carefully before you test it on a working telephone line. If the symptoms persist, you will have to reevaluate them and narrow the problem to another part of the equipment. If normal operation is restored or greatly improved, test the equipment's various functions. When you can verify that your symptoms have stopped during actual operation, the equipment can be returned to service.

Do not be discouraged if the equipment still malfunctions. Simply walk away, clear your head, and start again by defining your current symptoms. Never continue with a repair if you are tired or frustrated—tomorrow is another day. You should also realize that there might be more than one bad part to deal with. Remember that telephone equipment is just a collection of assemblies, and each assembly is a collection of parts. Normally, everything works together, but when one part fails, it can cause one or more interconnected parts to fail as well. Be prepared to make several attempts before the telephone or answering machine is repaired completely.

Notes on technical information

Information is perhaps your most valuable tool in tracking a repair of telephone equipment. Involved electronic troubleshooting generally requires a complete set of schematics and a parts list. Luckily, many manufacturers of telephone equipment do sell technical information for their products (or at least their older products). Sony and Tandy (Radio Shack) are just two manufacturers that make their technical data available to consumers. Contact the literature department of your particular manufacturer for specific data prices and availability; be sure to request a *service manual* (not an owner's or user's manual). If you are able to obtain technical information, it is strongly recommended that you have the data on hand before starting your repair. Service manuals often contain important information on custom or application-specific integrated circuits (ASICs) used in the equipment that you won't be able to obtain elsewhere.

Static electricity

With any type of electronic troubleshooting activity, there is always a risk of further damage being caused to equipment accidentally during the repair process. With sophisticated telephone electronics, that damage hazard comes in the form of electrostatic discharge (ESD) that can destroy sensitive electronic parts.

If you have ever walked across a carpeted floor on a cold, dry winter day, you probably experienced the effects of ESD firsthand while reaching for a metal object, such as a door knob. Under the right conditions, your body can accumulate static charge potentials greater than 20,000 V! When you provide a conductive path for electrons to flow, that built-up charge rushes away from your body at the point closest to the metal object. The result is often a brief, stinging shock. Such a jolt can be startling and annoying, but it is generally harmless to people. Semiconductor devices, on the other hand, are highly susceptible to real physical damage from ESD when you handle or replace circuit boards or integrated circuits. This section introduces you to static electricity and shows you how to prevent ESD damage during your repairs.

Static formation

When two dissimilar materials are rubbed together (such as a carpet and the soles of your shoes), the force of friction causes electrons to move from one material to another. The excess (or lack) of electrons causes a charge of equal but opposite polarities to develop on each material. Because electrons are not flowing, there is no current, so the charge is said to be *static*. However, the charge does exhibit a voltage potential. As materials continue to rub together, their charge increases—sometimes to potentials of thousands of volts.

In a human, static charges are often developed by normal, everyday activities such as combing your hair. Friction between the comb and your hair causes opposing charges to develop. Sliding across a vinyl car seat, pulling a sweater on or off, or taking clothes out of a dryer are just some of the ways static charges can appear in the body—it is virtually impossible to avoid. ESD is more pronounced in winter months when dry air works to allow a greater accumulation of charges. In the summer, humidity in the air tends to bleed away (or *short-circuit*) most accumulated charges before they reach shock levels that you can physically feel. Regardless of the season, though, ESD is always present to some degree—and always a danger to sensitive electronics.

Device damage

ESD poses a serious threat to most advanced ICs. They can easily be destroyed with static discharge levels of just a few hundred volts—well below your body's ability to feel a static discharge. Static discharge at sufficient levels can damage bipolar transistors, transistor-transistor logic (TTL) gates, emitter-coupled logic (ECL) gates, operational amplifiers (op-amps), silicon-controlled rectifiers (SCRs), and junction field-effect transistors (JFETs)—but certainly the most susceptible components to ESD are those ICs fabricated using metal-oxide semiconductor (MOS) technology. A typical MOS transistor is shown in Fig. 3-3.

The MOS family of devices (PMOS, NMOS, HMOS, CMOS, etc.) has become the cornerstone of high-performance ICs. They are used as memories, high-speed logic and microprocessors, and other advanced applications found in today's intelligent telephone equipment. Typical MOS ICs can easily fit over one million transistors onto a single IC die. Every part of these transistors must be made continually smaller to keep pace with the constant demand for ever-higher levels of IC complexity. As each part of the transistor shrinks, however, their inherent breakdown voltage drops, and their susceptibility to ESD damage escalates.

A typical MOS transistor breakdown is illustrated in Fig. 3-4. Notice the areas of positive and negative semiconductor material that forms its three terminals: source, gate, and drain. The *gate* is isolated from the other parts of the transistor by a thin film of silicon dioxide, sometimes called the "oxide layer." Unfortunately, this layer is extremely thin. High voltages, like those voltages from electrostatic discharges, can easily overload the oxide layer—this results in a permanent puncture through the gate. Once this happens, the transistor—and therefore the entire IC—is permanently defective and must be replaced.

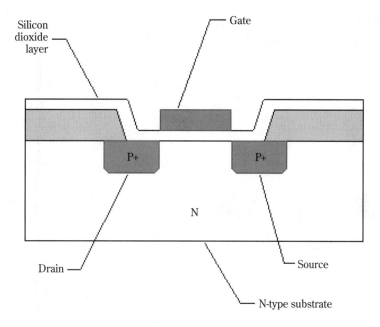

3-3 A typical MOS transistor.

Controlling static electricity

Never underestimate the importance of static control during your repairs. Without realizing it, you could destroy a new IC or circuit board before you even have the chance to install it—and you would never even know that static damage has occurred. All it takes is the careless touch of a charged hand or a loose piece of clothing. Take the necessary steps to ensure the safe handling and replacement of your sensitive (and expensive) electronics.

One way to control static is to keep charges away from boards and ICs to begin with. This is often accomplished as part of a device's packaging and shipping container. ICs are typically packed in a specially-made conductive foam. Carbon granules are compounded right into the polyethylene foam to achieve conductivity (about 3000 ohms-per-centimeter). Foam support helps to resist IC lead bending, absorb vibrations, and keeps every lead of the IC at the same potential (known as *equipotential bonding*). Conductive foam is reusable, so you can insert ICs for safe-keeping, then remove them as needed. You can purchase conductive foam from just about any retail electronics store.

Circuit boards are normally held in conductive plastic bags that dissipate static charges before damage can occur. Anti-static bags are made up of different material layers—each material exhibiting different amounts of conductivity. The bag acts as a *Faraday cage* for the device it contains. Electrons from an ESD will dissipate along a bag's surface layers instead of passing through the bag wall to its contents. Anti-static bags are also available through many retail electronics stores.

Whenever you work with sensitive electronics, it is a good idea to dissipate charges that might have accumulated on your body. A conductive fabric wrist strap

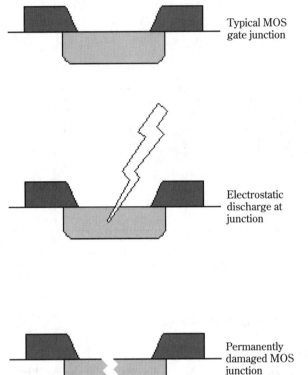

Typical MOS
gate junction

Electrostatic
discharge at
junction

3-4 ESD breakdown in a
typical MOS transistor.

Permanently
damaged MOS
junction

that is soundly connected to an earth ground will slowly bleed any charges away from your body. Avoid grabbing hold of a ground directly. Although this will discharge you, it can result in a nasty jolt if you have picked up a large electrostatic charge.

Remember to make careful use of your static controls. Keep ICs and circuit boards in their anti-static containers at all times. Never place parts onto or into synthetic materials (such as nonconductive plastic cabinets or fabric coverings) that could hold a charge. Handle static-sensitive parts carefully. Avoid touching IC pins if at all possible. Be sure to use a conductive wrist strap connected to a reliable earth ground.

Soldering

Soldering is the most commonly used method of connecting wires and components within an electrical or electronic circuit. Metal surfaces (in this case, component leads, wires, or printed circuit boards) are heated to high temperatures, then joined together with a layer of compatible metal in its molten state. When performed correctly with the right materials, soldering forms a lasting, corrosion-proof, inter-molecular bond that is mechanically strong and electrically sound. All that is required is the appropriate soldering iron and electronics-grade (60/40) solder. This section explains the tools and techniques for both regular and surface mount soldering.

Soldering background

By strict definition, *soldering* is a process of bonding metals together. There are three distinct types of soldering: brazing, silver soldering, and soft soldering. Brazing and silver soldering are used when working with hard or precious metals, but soft soldering is the technique of choice for electronics work.

In order to bond wires or component leads (typically made of copper), a third metal must be added while in its molten state. The bonding metal is known simply as *solder*. Several different types of solder are available to handle each soldering technique, but the chosen solder must be compatible with the metals to be bonded—otherwise, a bond will not form.

Lead and tin are two common, inexpensive metals that adhere very well to copper. However, neither metal by itself has the strength, hardness, and melting point characteristics to make it practically useful. Therefore, lead and tin are combined into an alloy. A ratio of approximately 60% tin and 40% lead yields an alloy that offers reasonable hardness, good pliability, and a relatively low melting point that is ideal for electronics work. This is the solder that must be used.

While solder adheres very well to copper, it does not adhere well at all to the natural oxides that form on a conductor's surface. Even though conductors might look clean with the unaided eye, some amount of oxidation is always present. Oxides must be removed before a good bond can be achieved. A resin cleaning agent (called *flux*) can be applied to conductors before soldering. Resin is chemically inactive at room temperature, but it becomes extremely active when heated to soldering temperatures. Activated flux bonds with oxides and strips them away from the copper surface. As a completed solder joint cools, residual resin also cools and returns safely to an inactive state.

Never use an acid or solvent-based flux to prepare conductors. Acid fluxes can clean away oxides as well as resin, but acids and solvents remain active after the joint cools. Over time, active acid flux will dissolve copper wires and eventually cause circuit failure. Resin flux can be purchased as a paste that can be brushed onto conductors before soldering, but most electronic solders have a core of resin manufactured right into the solder strand itself. Prefabricated flux eliminates the mess of flux paste, and cleans the joint as solder is applied.

Irons and tips

A soldering iron is little more than a resistive heating element built into the end of a long steel tube as shown in the cross-sectional diagram of Fig. 3-5. When 120 Vac is applied to the heater, it warms the base of a metal tip. Any heat conducted down the cooldown tube (toward the handle) is dissipated harmlessly to the surrounding air. This keeps the handle temperature low enough to hold comfortably.

Although some heat is wasted along the cooldown tube, most of the heat is channeled into a soldering tip similar to the one shown in Fig. 3-6. Tips generally have a core of solid copper that is plated with iron. The plated core is then coated with a layer of nickel to stop high-temperature metal corrosion. The entire assembly, except for the tip's very end, is finally plated with chromium—it gives a new tip its shiny chrome appearance. A chromium coating renders the tip *non-wettable*—solder will not stick to it. Because solder must stick at the tip's end, that end is plated with tin. A tin coating (a basic component of solder) makes the tip *wettable* so that molten solder

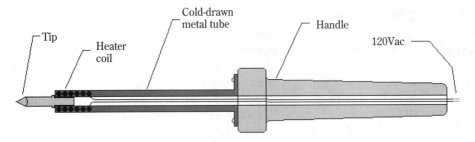

3-5   A cross section of a simple soldering iron.

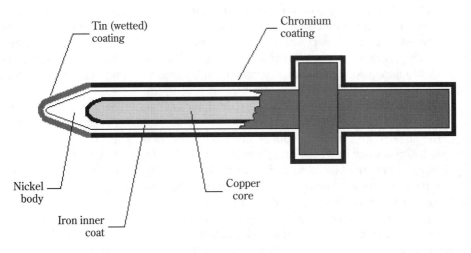

3-6   A typical soldering tip.

will adhere. Tips are manufactured in a wide variety of shapes and sizes to handle different soldering tasks. Before you can select the best tip for the job, you must understand ideal soldering conditions.

The very best soldering connections are made within only a narrow window of time and temperature. A solder joint heated between 500 to 550° F for 1 to 2 seconds will make the best connections. You should select a soldering iron wattage and tip shape to achieve these conditions. The purpose of soldering irons is not to melt solder—instead, a soldering iron is supposed to deliver heat to a joint, the joint should melt the solder.

A large solder joint (with larger or more numerous connections) requires a larger iron and tip than a small joint (with fewer or smaller connections). If you use a small iron to heat a large joint, the joint might dissipate heat faster than the iron can deliver it, so the joint might not reach an acceptable soldering temperature. Conversely, using a large iron to heat a small joint will overheat the joint. Overheating can melt wire insulation and damage printed circuit board traces. Match wattage to the application. Most general-purpose electronics work can be accomplished using a 25-to-30-watt soldering iron.

Because the end of a tip contacts the joint to be soldered, the tip's shape and size can greatly assist heat transfer. When heat must be applied across a wide area (such as a wire splice), a wide area tip should be used. A screwdriver (or *flat-blade*) tip such as shown in Fig. 3-7 is a good choice. If heat must be directed with pinpoint accuracy for small, tight joints or printed circuits, a narrow blade or conical tip is best. Two tips for surface mount desoldering are also shown in Fig 3-7. More on surface mount soldering later in this chapter.

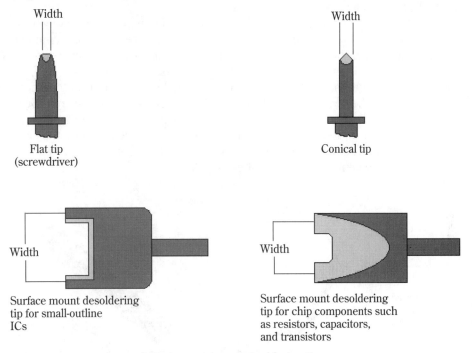

3-7   An assortment of soldering tips.

Soldering

Always keep your soldering iron parked in a secure holder while it is on. Never allow a hot iron to sit freely on a table top or anything that might be flammable. Make it a rule to always wear safety glasses when soldering. Active resin or molten solder can easily flick off the iron or joint and do permanent damage to the tissue in your eyes.

Give your iron plenty of time to warm up. Five minutes of warming time is usually adequate, but small-wattage irons (or irons with large tips) might need even more time. Once the iron is at its working temperature, you should coat the wettable portion of the tip with a layer of fresh solder—a process known as *tinning* the iron. Rub the tip into a sponge soaked in clean water to wipe away any accumulations of debris and carbon that might have formed, then apply a thin coating of fresh solder to the tip's end. Solder penetrates the tip to a molecular level and forms a cushion of molten solder that aids in heat transfer. Re-tin the iron whenever its tip becomes blackened—perhaps every few minutes, or after several joints.

It can be helpful to tin each individual conductor before actually making the complete joint. To tin a wire, prepare it by stripping away 3/16 to 1/4 inch of insulation. As you strip insulation, be sure not to nick or damage the conductor. Heat the exposed copper for about 1 second, then apply solder into the wire—not into the iron. If the iron and tip are appropriate, solder should flow evenly and smoothly into the conductor. Apply enough solder to bond each of a stranded wire's exposed strands. When tinning a solid wire or component lead, apply just enough solder to lightly coat the conductor's surface. You will find that conductors heat faster and solder flows better when all parts of a joint are tinned in advance.

Making a complete solder joint is just as easy. Bring together each of your conductors as necessary to form the joint. For example, if you are soldering a component into a printed circuit board, insert the component's leads into their appropriate PC board holes. Place the iron against all conductors to be heated as shown in Fig. 3-8. For a printed circuit board, heat the printed trace and component lead together. After about 1 second, flow solder gently into the hot conductors—not the iron. Be sure that solder flows cleanly and evenly into the joint. Apply solder for another 1 or 2 seconds, then remove both solder and iron. Do not attempt to touch or move the joint for several seconds. Wait until the solder cools and sets. If the joint requires additional solder, reheat the joint and flow in a bit more solder.

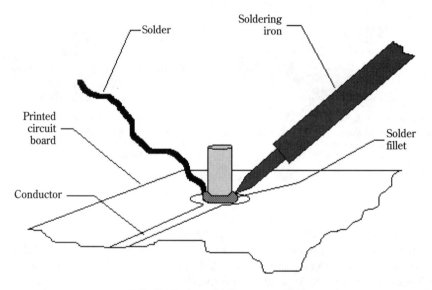

3-8   Soldering a basic connection.

You can identify a good solder joint by its smooth, even, silvery-grey appearance. Any charred or carbonized flux on the joint indicates that your soldering temperature was too high (or heat was applied for too long). Remember that solder cannot flow unless the joint is hot. If the joint is not hot, solder will cool before it bonds. The result is a rough, built-up, dull grey or blackish mound that does not adhere to the joint very well. This is known as a *cold* solder joint. A cold joint can often be corrected by reheating the joint properly and applying fresh solder.

Surface mount soldering

Conventional printed circuits use *through-hole* components. Parts are inserted on one side of the PC board, and their leads are soldered to printed circuit traces on the other side. Surface mounted (SM) components do not penetrate a printed circuit board. Instead, SM components rest on one side of a PC board as shown in Fig. 3-9. Metal tabs are used instead of long component leads. A full range of parts—from resistors and capacitors to transistors and ICs—are currently available in SM packages. Even advanced ICs like microprocessors and ASICs can be found in surface mount packages. Many surface mount PC assemblies are soldered using one of two soldering techniques: *flow* soldering, or *reflow* soldering.

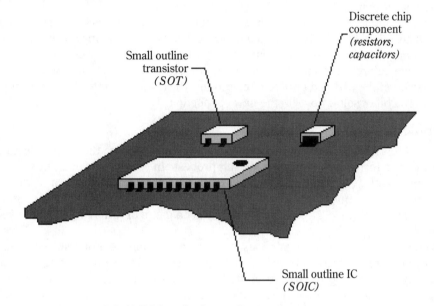

Discrete chip component *(resistors, capacitors)*

Small outline transistor *(SOT)*

Small outline IC *(SOIC)*

3-9   A PC board using surface mount components.

In the flow soldering technique, surface mount components are first glued into place on a PC board using automated assembly equipment. Gluing keeps each part in place and aligned with the proper PC board traces. The loaded PC board is then placed into a heating chamber and raised to soldering temperature. Molten solder is allowed to flow freely over the board where it adheres to heated component leads and PC board traces. This effectively solders the components into place. The remainder of the board is physically and chemically masked prior to soldering to prevent solder from sticking elsewhere.

A finished PC board is then cooled slowly to prevent excessive thermal shock to its components, the masks are stripped away, and the board can be tested (or any through-hole parts can be added). This type of fabrication is similar in principle to the techniques used to mass-produce conventional through-hole PC boards.

The technique of reflow soldering uses a slightly different approach. Before any components are installed, a masked PC board is coated with a paste form of solder. The board is heated to soldering temperature that allows solder to adhere to each PC trace. This forms a little bead of solder under each component lead. Any masking is stripped away. Components are then glued into place using automatic assembly equipment (it would take a very long time for a person to do that by eye), and the board is quickly reheated to soldering temperature. Solder melts again and reflows to adhere at each component terminal. Reflow establishes the component's connections. After slow, careful cooling, the board can be tested or through-hole components can be added.

While the specifics of each SM soldering technique will have little (if any) impact on your troubleshooting, you should have some basic understanding of how surface mount components are assembled in order to disassemble them properly during your repair.

Resoldering a new surface mount component is just as easy as any conventional through-hole part, but SM parts are often too small to hold during the soldering process. You might find it helpful to secure the new part over its PC traces with a small drop of glue before soldering. After the glue is dry, apply a regular soldering iron tip to a junction of the part and printed circuit. Allow the joint to heat for 1 or 2 seconds, then apply a bit of solder to form the new connection.

Desoldering

Ideally, to desolder an electronic connection, you must remove the intermolecular bond that has formed during the soldering process. In reality, however, this is virtually impossible. The best that you can hope for is to remove enough solder to gently break the connection apart without destroying the component or damaging the associated PC trace. Desoldering is basically a game of removing as much solder as possible.

You will find that some connections are very easy to remove. For instance, a wire inserted into a printed circuit board might be removed easily just by heating the joint and gently withdrawing the wire from its hole once solder is molten. Use extreme caution when desoldering! Leads and wires under tension can spring free once solder is molten. A springing wire can launch a bead of hot solder into the air. Always wear safety glasses or goggles to protect your eyes from flying solder.

Desoldering other types of connections, such as through-hole components, is usually more difficult. When desoldering a part with more than one wire, it is virtually impossible to heat all of its leads simultaneously while withdrawing the part. As a result, it is necessary to remove as much solder as possible from each lead, then gently break each lead free as shown in Fig. 3-10. Grab hold of each lead with a pair of needle-nose pliers and wiggle the lead back and forth gently until it breaks free. Solder can be removed using regular desoldering tools, such as the solder vacuum or solder wick illustrated in Fig. 3-11.

Surface mount components are typically small and specialized. You should try to find a soldering tip that will exactly fit the desired part. Two simple SM soldering iron tips are shown in Fig. 3-7. With a tip that precisely fits the part you wish to desolder, you can heat all joints in the part simultaneously, then move the part clear in one quick motion. Residual solder can be cleaned up later with conventional desoldering tools. There are also special tips for desoldering a selection of surface mount IC packages. Instead of using fixed SM tips, you could use SM soldering tweezers as shown in Fig. 3-12. This is a specialized soldering iron that can be used to desolder and grasp a wide range of surface mount parts.

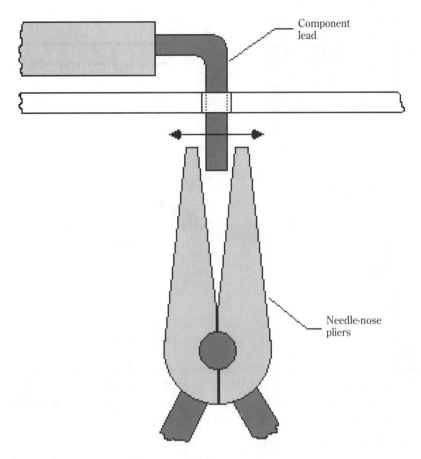

3-10 Breaking a stubborn solder connection.

Printed circuits

In the very early days of consumer electronics, circuit assemblies were manufactured by hand on bulky, metal frames. Each component was then wired together by hand. If you have ever seen a chassis from an old tube-driven television or radio, you have probably seen this type of construction. Eventually, the costs of hand-building electronic chassis became so high that a new technique was introduced—it used photographic processes to print wiring patterns onto copper-clad boards. Excess copper was then chemically stripped away, leaving only the desired wiring patterns. Parts could then be inserted into the board and soldered quickly, easily, and accurately.

Before long, manufacturers realized that these *printed circuits* appeared more uniform, were easier to inspect and test, required much less labor to assemble, and were lighter and less bulky than a metal chassis assembly. Today, virtually all electronic equipment, including your telephone or answering machine, incorporates

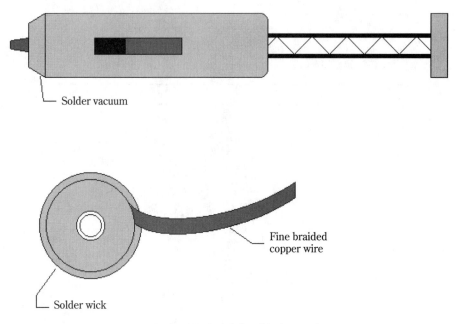

Solder vacuum

Fine braided
copper wire

Solder wick

3-11 Conventional desoldering tools.

some type of printed circuit board. The size and complexity of the board depends largely on the particular circuit's job. This section describes the major types of printed circuit boards that you might encounter, and presents a selection of PC board troubleshooting and repair techniques that you can use.

Types of printed circuits

Printed circuits are available as single-sided, double-sided, or multilayer boards. Each type of board can hold surface mount or through-hole components.

Single-sided PC boards are the simplest and least expensive type of printed circuit. Copper traces are etched on only one side of the board. Holes can then be

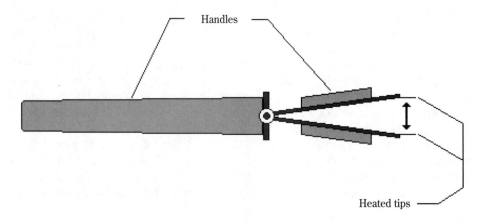

Handles

Heated tips

3-12 Soldering tweezers for surface mount desoldering.

drilled through the board to accommodate component leads. Components are inserted from the blank side of the board (the component side) so that their leads protrude on the copper trace side (the solder side). Component leads can then be soldered to their copper traces to complete the printed circuit.

When circuits become too complex to route all traces on one side of the PC board, traces can be etched onto both sides of a printed circuit. This is called a *double-sided* PC board. Plated (electrically conductive) holes are used to interconnect both sides of the board as needed. Such plated holes are also used to hold through-hole component leads. Solder conducts up the plated hole through capillary action and ensures that a component lead is properly connected to both sides of the board—this allows the board to be soldered from one side only during the manufacturing process. However, desoldering leads in plated holes can become somewhat difficult because solder adheres all the way through the hole. All internal solder must be removed before a lead can be withdrawn. If you pull a wire or component lead out before solder is removed, you stand a good chance of ripping the plating right out of the hole.

Even more complex circuit designs can be fabricated on *multilayer* PC boards. Not only will you find traces on both external sides of the PC board, but there can be even more layers of etched traces sandwiched between these two faces (each layer is separated by an insulating layer). As with double-sided boards, multilayer boards use plated through-holes to hold component leads, and bond various layers together.

Typical printed circuits use etched copper traces on a base material of paper-based phenolic or epoxy. Other printed circuits incorporate a base of glass-fabric epoxy, or some similar plastic-based substance. These types of materials offer a light, strong, rigid base for printed circuits.

A fourth (but less commonly used) type of printed circuit is known as the *flexible* printed circuit. Copper traces are deposited onto a layer of plastic such as Kapton. Traces can be included on both sides of this base layer to form a single or double-sided circuit. Traces are then covered by an insulating layer of plastic (often Kapton as well). Using alternate layers of copper traces and flexible insulation, it is possible to form multilayer flexible printed circuits.

Flexible circuits have the ability to fold and conform to tight or irregular spaces. As a result, flexible circuits are often used as wiring harnesses—that is, components are placed as needed, then a flexible circuit is inserted and attached (by screws or solder) to interconnect each component. Individual components are rarely soldered to a flexible PC as they are with a rigid PC.

Printed circuit repairs

Printed circuits are generally very reliable structures, but instances of physical abuse can easily damage the rigid phenolic or glass base, as well as any printed traces. If damage occurs to a PC board, you should know what signs of damage to look for, and what steps you can take to correct the damage in your telephone or answering machine. There are four general classes of PC board problems that you should know about: lead pull-through, printed trace break, board cracks, and heat damage.

Lead pull-through

Normally, a well-made solder joint will hold a wire or component lead tightly into its connection on a PC board. However, if that wire or lead is suddenly placed under a

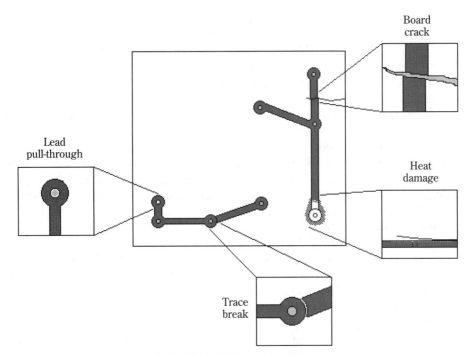

3-13 Typical PC board problems.

lot of stress, the solder joint can fail partially or completely as shown in Fig. 3-13. Stress can be applied with sudden, sharp movements such as dropping or striking the telephone.

Lead pull-through failure is not always obvious unless the lead or wire is out of its through-hole entirely. If the wire is still making contact with the PC board, its electrical connection might be broken or intermittent. You can test an intermittent connection by exposing the PC board, then gently rapping on the board or suspect conductor. By tapping different areas of the board, it might be possible to focus on an intermittent connection in the area that is most sensitive to the tapping. You can also test suspected intermittents by gently wiggling wires or component leads. The conductor most sensitive to the touch is probably the one that is intermittent.

Another case of lead pull-through can occur on double-sided or multilayer PC boards during desoldering. Various layers are connected together by plated holes. Component leads are typically soldered into plated through-holes. If you pull out a conductor without removing all the solder, you can rip out part or all of the hole's plating along with the conductor. When this happens, the electrical integrity at that point of the PC board is broken.

For double-sided PC boards, this can often be corrected by soldering the new component lead on both sides of the PC board. There is usually enough exposed copper on the component side to ensure a reasonable solder fillet. Unfortunately, there is no reliable way to solder a new lead to each layer of a multilayer board. As a result, a damaged through-hole on a multilayer board might be beyond repair.

The best way to avoid damaging a plated through-hole is to heat a joint while re-

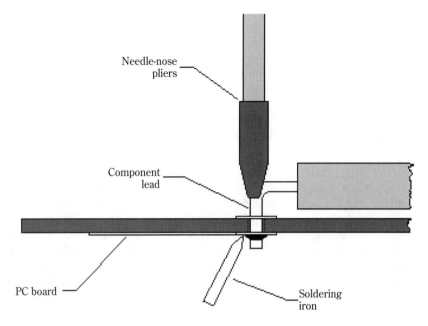

Needle-nose pliers

Component lead

PC board

Soldering iron

3-14 Ensuring a clean through-hole lead removal.

moving the lead simultaneously as shown in Fig. 3-14. Grasp the lead with a pair of needle-nose pliers while heating the joint. When solder is molten, pull out the component lead. You can then safely clean up any residual solder with conventional desoldering tools. Never grasp the component lead with your bare fingers. The entire lead reaches soldering temperatures almost immediately.

Printed trace break

Another common problem in printed circuits is *trace break*. This also can be the result of a physical shock or sudden impact to the PC board. In this case, a portion of the printed trace (usually where a solder pad meets the remainder of the trace) can suffer a fine, hairline fracture that results in an open or intermittent circuit, as illustrated in Fig. 3-13. What makes this especially difficult is that a trace break can be almost impossible to see on visual inspection. You must wiggle each solder pad until you find the fractured connection. Large, heavy components, such as transformers or relays, are prime candidates for trace breaks, so start your search there.

Do not attempt to create a bridge of solder across the break, or jumper directly across the fracture. Solder does not adhere well to the chemical coatings often used with PC boards, so such quick fixes rarely last long. To correct a printed trace break, you should desolder and remove the broken portion of the trace, then solder a jumper wire between two associated component leads. Do not solder directly to the printed trace.

Board cracks

Under extreme conditions, the phenolic or glass-epoxy circuit board itself can crack. This is not unusual for equipment that has been dropped or abused. When a crack occurs as shown in Fig. 3-13, the course of the crack might sever one or more printed traces. Luckily, board cracks are fairly easy to detect on sight. By following the crack, it is a simple matter to locate any severed traces.

The best, most reliable method of repairing broken traces is to solder a wire jumper between two associated solder pads or component leads. Never try to make a solder bridge across a break. Solder does not adhere well to the chemicals used on many PC board traces, so such repairs will not last long. If the physical crack is severe, you might want to work a bit of epoxy adhesive into the crack to help reinforce the board.

Heat damage

Printed copper traces are bonded firmly to the phenolic or glass epoxy board underneath. When extreme heat is applied to the copper traces, however, it is possible to separate the copper trace from the board. This type of damage usually occurs during soldering or desoldering when concentrated heat is applied with a soldering iron.

The only real remedy for this type of damage is to carefully cut off that portion of the separated trace (to prevent the loose copper from accidentally shorting out other components), and solder a wire jumper from the component lead to an adjacent solder pad or component lead.

4
Using test equipment

BEFORE YOU CAN TROUBLESHOOT AND REPAIR THE ELECTRONIC CIRCUITS IN your telephone or answering machine, you need some basic test equipment to measure such circuit parameters as voltage, current, and resistance—as well as more sophisticated parameters such as capacitance, frequency, and semiconductor junction conditions (Fig. 4-1). This chapter introduces you to the background, operations, and testing methods for three major test instruments: a multimeter, a logic probe, and an oscilloscope. It also explains the methods used to read standard component markings.

Multimeters

Multimeters are by far the handiest and most versatile pieces of test equipment you will ever use. If your toolbox does not contain a good-quality multimeter, now would be a good time to purchase one. Even the most basic multimeters are capable of measuring resistance, ac and dc voltage, and ac and dc current. For under $150, you can buy a digital multimeter that includes handy features like a capacitance checker, an extended current measuring range, a continuity buzzer, and even a diode and transistor checker. These features will aid you not only in telephone repairs, but in many other types of electronic repairs as well. Digital multimeters are easier to read, more tolerant of operator error, and more precise then their analog predecessors. Figure 4-2 illustrates a typical digital multimeter. Digital multimeters are referred to throughout this book.

There are usually just two considerations when setting up and using a multimeter. First, the meter must be set to the desired function (voltage, current, capacitance, etc.). Second, the meter's range of measurement must be set correctly for that

Photo provided courtesy of Code-A-Phone Corporation.

4-1   An integrated telephone/answering machine.

4-2   A digital multimeter. Courtesy of B+K Precision.

function. If you are unsure about what range to use, start by choosing the highest possible range. Once you actually make a measurement and get a better idea of what the actual reading will be, you can adjust the meter's range to achieve a more precise reading. If your signal exceeds the meter's range setting, an *overrange* warning displays. Many digital voltmeters are capable of selecting the most precise range automatically once a signal is applied (this feature is called *autoranging*).

A multimeter can be used for two general types of testing: static and dynamic. *Dynamic* tests are made with power on and connected to the circuit under test, while *static* tests are made on unpowered circuits or components. Measurements like voltage, current, and frequency are dynamic tests, but most other tests—such as resistance/continuity, capacitance, or diode quality—are static tests. The following sections present a series of simple multimeter measurement techniques. If you are new to this type of test equipment, you will find the following sections to be a good tutorial.

Measuring voltage

Voltage measurements are the most fundamental dynamic tests in electronics. Multimeters can measure both dc voltages (your particular meter might be marked DCV or Vdc for this function) and ac voltages (probably marked ACV or Vac). It is important to remember that all voltage measurements are taken in parallel with the desired circuit or component. Never interrupt a circuit and attempt to measure voltage in series with other components. Any such reading would be meaningless and your circuit might not even function.

Set your multimeter to the appropriate function (DCV or ACV), then select the proper range for the voltages that you will be measuring. If you are unsure what range to use, start with the largest possible range. This helps to prevent accidental damage to the meter. Keep in mind that an autoranging multimeter will select its own range once a signal is applied. Place your test leads in parallel across the part under test as shown in Fig. 4-3, then read voltage directly from the digital display. Dc voltage readings are polarity sensitive, so if you read +5 Vdc and then reverse the test leads, you will see a reading of –5 Vdc. Ac voltage readings are not polarity sensitive.

Measuring current

Most general-purpose multimeters allow you to measure ac current (usually marked ACA or Iac) and dc current (marked DCA or Idc) in an operating circuit, although there are typically fewer ranges to choose from. As with voltage measurements, current is measured in a working circuit with power on, but current must be measured in series with the circuit or component under test.

Inserting a meter in series, however, is not always a simple task. In many cases, you must interrupt a circuit at the point you wish to measure, then connect your test leads across the break. While it can be quite easy to interrupt a circuit, remember that you must also put the circuit back together, so use care when choosing a point to break. Never attempt to measure current in parallel across a component or circuit. Current meters, by nature, exhibit a very low resistance across their test leads—often below 0.1 Ω. Placing a current meter in parallel can cause a short circuit across a component—that can damage that component, the circuit under test, or the meter itself.

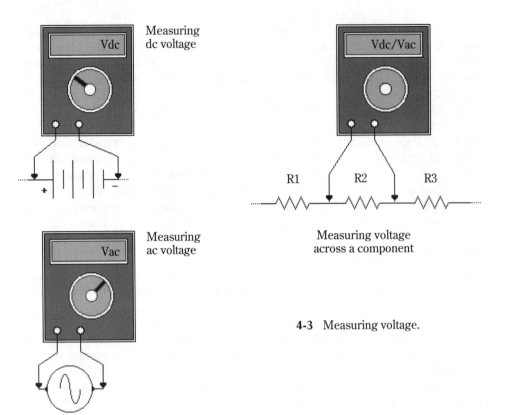

4-3   Measuring voltage.

Set your multimeter to the desired function (DCA or ACA) and select the appropriate range. If you are unsure about the proper range, set the meter to its largest range. It is usually necessary to plug one of your test leads into a different *current input* jack on the multimeter. Unless your multimeter is protected by an internal fuse (most meters are protected), its internal current measurement circuits can be damaged by excessive current. Make sure that your meter can handle the maximum amount of current you expect.

Turn off all power to a circuit before inserting a current meter. Deactivation prevents any unpredictable circuit operation when you actually interrupt it. If you want to measure the power supply current feeding a circuit such as in Fig. 4-4, break the power supply line at any convenient point, insert the meter carefully, then reapply power. Read current directly from the meter's display. This procedure can also be used for taking current measurements within a circuit.

Measuring frequency

Some multimeters offer a frequency counter (marked f or Hz) that can read the frequency of a signal. The ranges that are available will depend on your particular meter. Simple hand-held meters can often read up to 100 kHz, while benchtop models

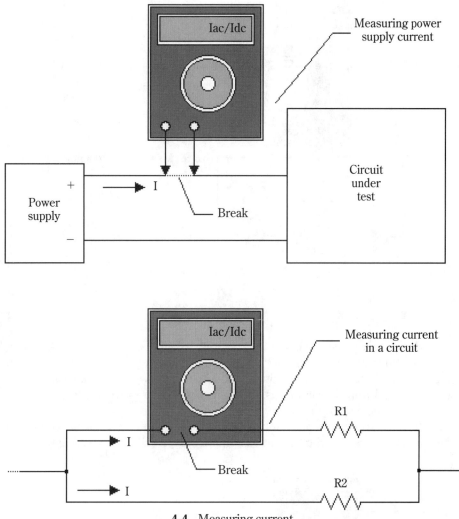

4-4 Measuring current.

can handle 10 MHz or more. Frequency measurements are dynamic readings made with circuit power applied.

Set your multimeter to its frequency counter function and select the appropriate range. If you are unsure just what range to use, start your measurements at the highest possible range. Place your test leads in parallel across the component or circuit to be tested as shown in Fig. 4-5, and read frequency directly from the meter's display. An autoranging multimeter selects the proper range after the signal is applied.

Measuring resistance

Resistance (ohms) is the most common static measurement that your multimeter is capable of. This is a handy function, not only for checking resistors themselves, but

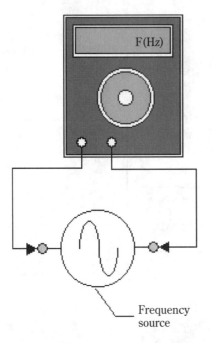

4-5 Measuring frequency.

for checking other resistive elements like wires, solenoids, motors, connectors, and some basic semiconductor components. Resistance is a static test, so all power to the component or circuit must be removed. It might be necessary to remove at least one component lead from its circuit, as shown in Fig. 4-6, to prevent interconnections with other components from causing false readings.

Ordinary resistors can be checked simply by switching to a resistance function (often marked ohms, or with the Greek symbol omega (Ω) and selecting the appro-

Disconnect at least one component lead from the circuit to prevent false resistance readings

4-6 Measuring resistance.

priate range. Autoranging multimeters select the proper range after the meter's test leads are connected. Many multimeters can reliably measure resistance up to about 20 MΩ. Place your test leads in parallel across the component and read resistance directly from the meter's display. If resistance exceeds the selected range, the display will indicate an overrange (or infinite resistance) connection.

Continuity checks are made to ensure a reliable, low-resistance connection between two points. For example, you could check the continuity of a cable between two connectors to ensure that both ends are connected properly. Set your multimeter to a low resistance scale, then place your test leads across both points to measure, as shown in Fig. 4-7. Ideally, good continuity should be about 0 Ω.

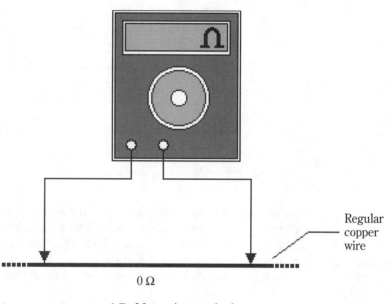

Regular copper wire

0 Ω

4-7 Measuring continuity.

Reading resistors

Conventional carbon resistors use a system of color codes to mark their value as shown in Fig. 4-8. When a resistor is held so that the color band closest to the end of the resistor is on the left as in Fig. 4-8, the resistor's value is interpreted from left to right. The first three color bands are used to specify a value. Bands 1 and 2 specify the amount, while band 3 specifies the multiplier.

For example, suppose a resistor offered the color sequence: brown, black, red. You can see from the table in Fig. 4-8 that brown = 1, black = 0, and red = 2 (red is in the multiplier position, so its value would be read as 100 Ω). The sequence of brown, black, and red would be read as 10 × 100, or 1000 Ω (or 1 kΩ). If the first three color bands of a resistor read: red, red, orange, the resistor would be read as 22 × 1000 (red, red, × orange) or 22,000 Ω (22 kΩ), and so on. Typical resistors can range from fractions of an ohm to 20 MΩ or more.

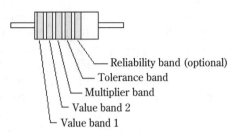

Color	Value Band 1	Value Band 2	Multiplier	Tolerance
Black	0	0	1	
Brown	1	1	10	
Red	2	2	100	
Orange	3	3	1000	
Yellow	4	4	10000	
Green	5	5	100000	
Blue	6	6	1000000	
Violet	7	7	10000000	
Gray	8	8	100000000	
White	9	9	- - - - - - - - -	
Gold	-	-	- - - - - - - - -	± 5%
Silver	-	-	- - - - - - - - -	± 10%
None	-	-	- - - - - - - - -	± 20%

4-8   Reading a resistor's color code.

Some resistors show a fourth band. This is known as the *tolerance band.* A gold band represents a tolerance of ±5%, and a silver band shows a tolerance of ±10%. If there is no tolerance band, the resistor is assumed to have a tolerance of ±20%. The tolerance band indicates just how close a resistor's actual value must be to its marked value. A low-tolerance resistor is closer to its marked value than a high-tolerance resistor.

On rare occasions, you might encounter a resistor with a fifth band. This is a *reliability band* often found on military-grade components. You will rarely encounter a fifth band, but if you do, you can ignore it for all practical purposes.

Surface mount resistors use a rather cryptic method of numerical marking to indicate values as shown in Fig. 4-9. The first two numbers on the component indicate its value, while the third number represents the multiplier. For the surface mount resistor in Fig. 4-9, it would be read as 10×100 (add two zeros) or $1000 \ \Omega$.

Checking a capacitor

There are two methods of checking a capacitor using your multimeter. If your multimeter is equipped with a built-in capacitance checker, all you need to do is select the capacitance function, set the desired range of capacitance to be measured, then connect your test probes in parallel across the capacitor under test. You might

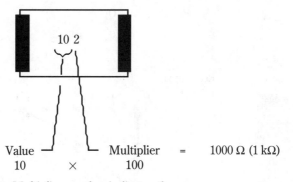

Value ⎤   ⎡ Multiplier   =   1000 Ω (1 kΩ)
  10   ×   100

Multiplier number indicates the
number of zeros to add after the value
digits

4-9   Reading surface mount resistor markings.

have to remove at least one of the component's leads to prevent its interconnections with other components from adversely affecting the meter's readings. In
some cases, you must remove the capacitor entirely and place it into a special test
fixture on the meter's face.

A capacitance meter usually displays the capacitance value directly in microfarads (μF) or picofarads (pF). As long as your reading falls within the tolerance of
the capacitor's marked value, you know the part is good. General-purpose capacitors
are rated at ±20% tolerance.

If your multimeter is not equipped with an internal capacitor checker, you can
still use the resistance ranges of your ohmmeter function to approximate a capacitor's quality. This type of check, as described below, provides a "quick and dirty"
judgement of whether the capacitor is good or bad. The principle behind this type of
check is simple—all ohmmeter ranges use an internal battery to supply current to
the component under test. When that current is applied to a working capacitor as
shown in Fig. 4-10, it causes the capacitor to charge. Charge accumulates as the ohmmeter is left connected. When first connected, the uncharged capacitor draws a
healthy amount of current—this reads as low resistance. As the capacitor charges, its
rate of charge slows down and less and less current is drawn as time goes on—this
results in a gradually increasing resistance level. Ideally, a fully charged capacitor
will stop drawing current—this results in an overrange or infinite resistance display.
When a capacitor behaves in this way, it is probably good.

Understand that you are not actually measuring resistance or capacitance here,
but only the profile of a capacitor's charging characteristic. If the capacitor is extremely small, or is open-circuited, it will not accept any substantial charge, so the
multimeter will read infinity almost immediately. If a capacitor is partially (or totally)
short circuited, it will not hold a charge, so you might read zero ohms (or resistance

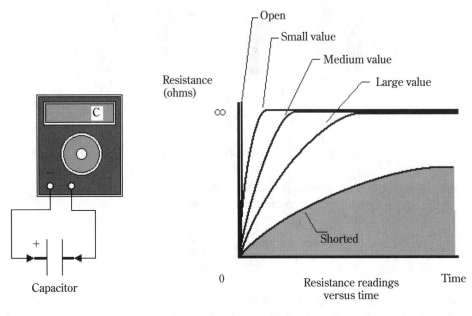

4-10   Measuring the quality of a capacitor.

can climb to some value below infinity and remain there). In either case, the capacitor is probably defective. If you doubt your readings, check several other capacitors of the same value and compare readings. Be sure to make this test on a moderate to high resistance scale. A low resistance scale can overrange too quickly to achieve a clear reading.

Reading capacitors

You will find that capacitors are typically marked in one of three ways. Many capacitors are plainly marked with their capacitance value (such as 100 µF, 0.22 µF, etc.). This is certainly the most common and accepted marking method among physically large ceramic and electrolytic capacitors that have enough surface area to hold these markings.

Extremely small capacitors, and many capacitors manufactured outside of the United States, use a numerical code to represent the capacitor's value. Figure 4-11 represents such techniques. For the ceramic capacitor marked 104, the first two digits are the value (or *significant*) digits, while the multiplier digit moves the decimal place that number of places to the left. The marking 104 would mean the number 10 with the decimal moved four places left, so the capacitor would be 0.0010 µF (or 0.001 µF). For the surface mount capacitor marked 331, the first two digits are also significant digits, but the multiplier moves the decimal place to the right, so 331 would actually be read as 330 pF. The number 332 would be read as 3300 pF, and so on. Whether the decimal is moved left or right depends largely upon the particular manufacturer, but you can always check the part with a capacitance checker just to make sure.

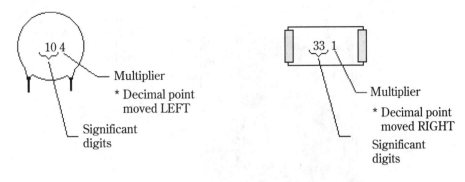

10. = 0.0010 = 0.001 µF   33 = 330. = 330 pF

* Verify the value of a capacitor using a capacitance checker

4-11   Reading capacitor markings.

Semiconductor checks

Many multimeters have a semiconductor junction checker for diodes and bipolar transistors. Meters equipped with a diode range in their resistance function can be used to measure the static resistance of most common diodes in their forward or reverse biased conditions, as shown in Fig. 4-12.

Before making any diode measurement, be sure that at least one of the diode's leads are removed from the circuit. This kind of isolation prevents other interconnections from causing false readings. Select the diode range from your multimeter's resistance function, and place your test leads in parallel across the diode in the forward bias direction. A working silicon diode should exhibit a static resistance between about 450 and 700 Ω directly on the meter's display. Reverse the orientation of your test probes to reverse bias the diode. Because a working diode will not conduct at all in the reverse direction, you should read infinite resistance.

A short-circuited diode exhibits a very low resistance in both the forward and reverse biased directions. This indicates a shorted semiconductor junction. An open-circuited diode exhibits very high resistance (usually infinity) in both its forward and reverse biased directions. A diode that is opened or shorted must be replaced. If you aren't sure how to interpret your measurements, test several other comparable diodes and compare readings.

Transistors are slightly more sophisticated semiconductor devices. They can be checked in several ways. Some multimeters feature a built-in transistor checker that measures a transistor's gain directly. If your meter offers a transistor checker, insert the transistor into the test fixture on the meter's face. Because transistors are three-terminal devices (emitter, base, collector), they must be inserted into the meter in their proper lead orientation before you can achieve a correct reading. Manufacturer's data sheets for a transistor will identify each lead and tell you the approximate gain reading that you should expect to see. An unusually low (or zero) reading indicates a shorted transistor, while a high (or infinite) reading suggests an opened transistor.

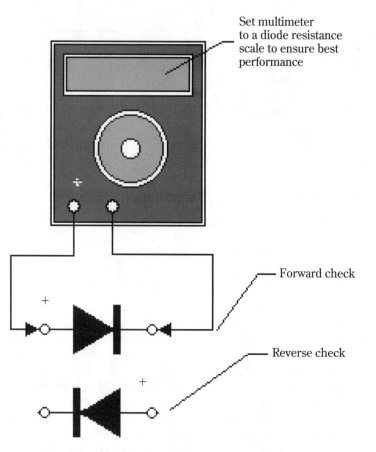

Set multimeter
to a diode resistance
scale to ensure best
performance

Forward check

Reverse check

4-12   Checking semiconductor diode junctions.

If your multimeter does not have a built-in transistor checker, its diode checker function can be used to inspect a transistor's base-emitter and base-collector junctions, as shown in Fig. 4-13. Each junction in a bipolar transistor acts like a semiconductor diode and yields similar readings. You might have to remove the transistor from its circuit so its interconnections with other components do not cause false readings.

Set your multimeter to its diode range, then place your test leads across the transistor's base-collector junction. If your particular transistor is the NPN-type (manufacturer's data or a corresponding schematic symbol will tell you), place the positive test lead at the base. This should forward-bias the base-collector junction and result in a normal amount of diode resistance (usually 450 to 700 Ω). Reverse your test leads across the base-collector junction. The junction should now be reverse biased and show infinite resistance. Repeat this entire procedure for the base-emitter junction.

If your bipolar transistor is the PNP-type, placement of the test leads will have to be reversed. For example, a forward-biased junction in an NPN transistor is reverse

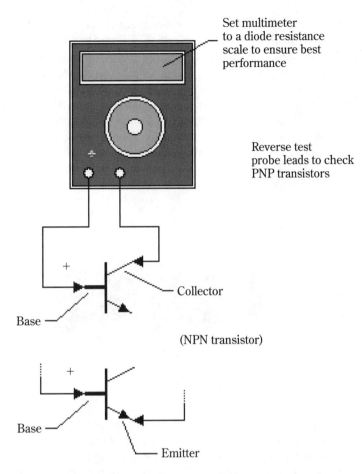

Set multimeter
to a diode resistance
scale to ensure best
performance

Reverse test
probe leads to check
PNP transistors

Collector

Base

(NPN transistor)

Base

Emitter

4-13 Checking semiconductor transistor junctions.

biased in a PNP transistor. Even though the three signal leads appear to be in the same location, the internal construction of NPN and PNP transistors is reversed.

As a final check, measure the diode resistance from emitter to collector directly. This test can be performed with NPN and PNP transistors. You should read infinite resistance in both test lead orientations. Although there is no diode junction from emitter to collector, short circuits can sometimes develop during a transistor failure.

Logic probes

The problem with most multimeters is that they do not relate very well to digital logic circuits. A multimeter can certainly measure whether a logic voltage is on or off, but if that logic signal changes quickly, the dc voltmeter function will not be able to track it properly—if at all. Logic probes are little more than extremely simple voltage sensors, but they provide a fast and easy means of detecting steady-state or alternating logic levels when you are working with digital logic ICs. Some logic probes can detect digital clock pulses or logic signals operating at speeds greater than 50 MHz.

Logic probes are rather simple-looking devices, as shown in Fig. 4-14. Indeed, it might be the simplest test instrument you ever use. A logic probe can be powered from its own internal battery, or from the circuit under test. Regardless of how the probe is powered, it must be connected into the common ground of the circuit being tested to ensure a common reference level. If the probe is powered from the circuit under test, attach the probe's power lead to a logic supply voltage source in the circuit.

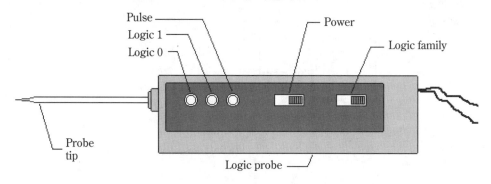

4-14 A basic logic probe.

Many logic probes use a series of LED indicators to display the measured logic state; a *logic high* (or 1), a *logic low* (or 0), or a *pulse* (or clock) signal. Some models of logic probe offer a switch that allows the probe to operate with two common logic families (transistor-transistor logic—TTL, or complementary metal-oxide semiconductor—CMOS). You will sometimes find TTL and CMOS devices mixed in the same circuit, but one family of logic devices will usually dominate.

In order to use a logic probe, touch its metal tip to the desired IC or component lead. Be certain that the point you wish to measure is, in fact, a logic point—high voltage signals can damage your logic probe. The logic state is interpreted by a few simple gates within the probe, then displayed on the appropriate LED. Logic probes are most useful for troubleshooting working logic circuits where logic levels and clock signals must be determined quickly and simply.

Oscilloscopes

Oscilloscopes offer a tremendous advantage over multimeters and logic probes. Instead of reading signals in terms of numbers or lighted indicators, oscilloscopes show voltage versus time on a graphic cathode-ray tube (CRT) display. Not only can you observe ac and dc voltages, but oscilloscopes enable you to watch voltages, or any other unusual signals, occur in real time. If you have used an oscilloscope—or seen one used—then you probably know just how useful they can be. Oscilloscopes,

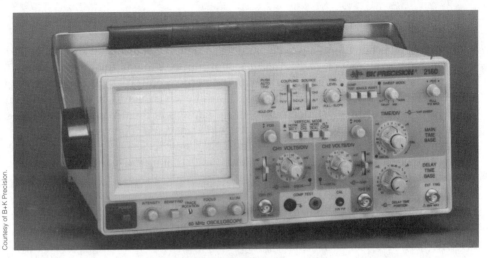

4-15 A 60-MHz oscilloscope.

such as the one shown in Fig. 4-15, might appear somewhat overwhelming at first, but many of their operations work the same way regardless of the model.

Controls

In spite of their wide variations in features and complexity, most controls are common to the operation of every oscilloscope. Controls fall into four categories: horizontal (time base) control, vertical (voltage sensitivity) control, housekeeping controls, and optional (enhanced) controls.

Housekeeping controls handle such mundane functions as oscilloscope power, trace intensity, graticule intensity, trace magnification, horizontal trace offset, vertical trace offset, and the trace finder. Generally speaking, any control that affects the trace itself (not the way trace information is being displayed) can be called a housekeeping control.

Because an oscilloscope displays voltage versus time, adjusting either the voltage sensitivity or time base settings will alter the display's appearance. Horizontal controls manipulate the left-to-right time appearance (or sweep) of the voltage signal. Your oscilloscope's master time base is adjusted using the TIME/DIV knob or buttons. This sets the rate at which voltage signals are swept onto the screen. As a general rule, smaller TIME/DIV settings allow shorter events to be displayed more clearly, and vice versa.

Remaining horizontal controls include a horizontal display mode selector, sweep trigger selector and sensitivity, trigger coupling selection, and trigger source selection. Your particular oscilloscope might offer even more controls.

An adjustment to the oscilloscope's voltage sensitivity will also alter your display. Vertical controls affect the deflection (up-to-down) appearance of your signal. An oscilloscope's vertical sensitivity is controlled with the VOLTS/DIV knob or buttons. When the VOLTS/DIV setting is made smaller (voltage sensitivity is increased), signals will appear larger vertically. A larger VOLTS/DIV setting decreases voltage sensitivity, so voltage signals will appear smaller vertically. Other vertical controls include a coupling selection, vertical mode selection, and display inverter switch.

Your oscilloscope might have any number of optional controls depending on its cost and complexity, but *cursor* and *storage* controls are some of the most common. Many digital oscilloscopes have horizontal and vertical on-screen cursors to aid in the evaluation of waveforms. Panel controls permit each cursor to be moved around the screen. The relative distance between the cursors is then calculated and converted to a corresponding voltage, time, or frequency value. The value can then be displayed on the screen in appropriate units. Storage oscilloscopes allow a screen display to be held right on the screen, or saved in digital memory within the oscilloscope to be viewed at a later time.

Oscilloscope specifications

There are a variety of important specifications that you should be familiar with when selecting and using an oscilloscope. The first specification to know is called *bandwidth*. Bandwidth represents the absolute range of frequencies that an oscilloscope can work with. Bandwidth does not mean that all signals within that frequency range will be displayed clearly. Bandwidth is usually rated from dc to some maximum frequency. For example, a relatively inexpensive oscilloscope might cover dc to 20 MHz, while a top-of-the-line model can work up to 150 MHz or more. Broad bandwidth is very expensive—more so than any other feature.

The vertical deflection (or vertical sensitivity) is another important specification. It is listed as the minimum to maximum VOLTS/DIV settings that are offered, and the number of steps that are available within that range. A typical model can provide vertical sensitivity from 5 mV/DIV to 5 V/DIV broken down into ten steps.

A time base (or sweep range) specification represents the minimum to maximum time base rates that an oscilloscope can produce, and the number of TIME/DIV increments that are available within that range. A range of 0.1 μs/DIV to 0.2 s/DIV in twenty steps is not unusual. You will typically find a greater number of time base increments than sensitivity increments.

There is a *maximum voltage input* that an oscilloscope is capable of handling. If that input voltage level is exceeded, the oscilloscope's input amplifier can be damaged. A maximum voltage input of 400 V (dc or peak ac) is common for most basic models, but more sophisticated models can accept input voltages better than 1000 V. An oscilloscope's input will present a load to whatever circuit or component it is placed across. This is called *input impedance*, and is usually expressed as a combined value of resistance and capacitance. To guarantee proper operation over the model's entire bandwidth, select a probe with load characteristics similar to those of the oscilloscope. Most oscilloscopes have an input impedance of 1 MΩ and a capacitance anywhere from 10 to 50 pF.

The accuracy of an oscilloscope represents the vertical and horizontal accuracy presented in the CRT display. Oscilloscopes are usually not as accurate as multimeters. Typical oscilloscopes can provide ±3% accuracy, so a 1-V measurement is displayed somewhere between 0.97 V to 1.03 V. Keep in mind that this does not consider human errors in reading the CRT's graticule. The strength of an oscilloscope is its ability to graphically display complex and quickly changing signals, so a 3% accuracy rating is often more than adequate.

Oscilloscope start-up procedure

Before you begin taking measurements, a clear, stable trace must be obtained (if not already visible). If a trace is not already visible, first make sure that any CRT screen storage modes are turned off, and that trace intensity is turned up to at least 50%. Set trace triggering to its automatic mode and adjust the horizontal and vertical offset controls to the center of their ranges. Be sure to select an internal trigger source, then adjust the trigger level until a stable trace is displayed. Vary your vertical offset if necessary to center the trace in the CRT.

If a trace is not yet visible, use the *beam finder* to reveal the beam's location. A beam finder simply compresses the vertical and horizontal ranges to force the trace onto the display. This gives you a rough idea of the trace's relative position. Once you are able to finally move the trace into position, adjust your focus and intensity controls to obtain a crisp, sharp trace. Keep intensity at a moderately low level to improve display accuracy and preserve the phosphor coating in the CRT.

Your oscilloscope should be calibrated to its probe before use. Calibration is a quick and straightforward operation that requires only a low-amplitude, low-frequency square wave. Many models have a built-in calibration signal generator (usually a 1-kHz, 300-mV square wave with a duty cycle of 50%). Attach your probe to the desired input jack, then place it across the calibration signal. Adjust your horizontal (TIME/DIV) and vertical (VOLTS/DIV) controls so that one or two complete cycles are clearly shown on the CRT.

Observe the visual characteristics of the test signal as shown in Fig. 4-16. If the square wave's corners are rounded, there might not be enough probe capacitance. Spiked square wave corners suggest too much capacitance in the probe. Either way, the scope and probe are not matched properly. You must adjust the probe capacitance control (sometimes labeled Cprobe) to establish a good electrical match—otherwise, signal distortion will interfere with your measurements. Slowly adjust the variable capacitance of your probe until the corners shown on the calibration signal are as square as possible. If you are not able to achieve a clean square wave, try a different probe.

Voltage measurements

The first step in any voltage measurement is to set your normal trace (called the *baseline*) where you want it. Normally, the baseline is placed along the center of the graticule during start-up, but it can be placed anywhere along the CRT so long as the trace is visible. To establish a baseline, switch your input coupling control to its ground position. Grounding the input disconnects any existing input signal and ensures a zero reading. Adjust the vertical offset control to shift the baseline wherever you want the zero reading to be, usually in the display center. If you have no particular preference, simply center the trace in the CRT.

To measure dc, set your input coupling switch to its dc position, then adjust the VOLTS/DIV control to provide the desired amount of sensitivity. If you are unsure just

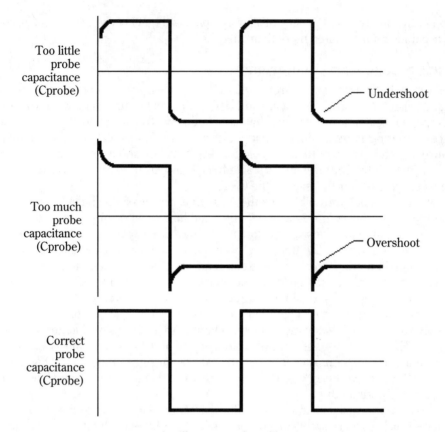

Too little
probe
capacitance
(Cprobe)

Undershoot

Too much
probe
capacitance
(Cprobe)

Overshoot

Correct
probe
capacitance
(Cprobe)

4-16   Oscilloscope probe calibration signals.

which sensitivity is appropriate, start with a very low sensitivity (a large VOLTS/DIV setting), then carefully increase sensitivity (gradually reduce the VOLTS/DIV setting) after your input signal is connected. This procedure prevents a trace from simply jumping off the screen when an unknown signal is first applied. If your signal does happen to leave the visible portion of the display, you could reduce the sensitivity (gradually increase the VOLTS/DIV setting) to make the trace visible again.

For example, suppose you were measuring a +5-Vdc power supply output. If VOLTS/DIV is set to 5 VOLTS/DIV, each major vertical division of the CRT display represents 5 Vdc, so your +5-Vdc signal should appear 1 full division above the baseline

$$5 \text{ VOLTS/DIV} \times 1 \text{ DIV} = 5 \text{ Vdc}$$

as shown in Fig. 4-17. At a VOLTS/DIV setting of 2 VOLTS/DIV, the same +5-Vdc signal would now appear 2.5 divisions above your baseline.

$$2 \text{ VOLTS/DIV} \times 2.5 \text{ DIV} = 5 \text{ Vdc}$$

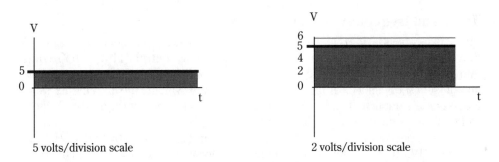

5 volts/division scale   2 volts/division scale

4-17 Reading dc voltages on an oscilloscope.

If your input signal was a negative voltage, the trace would appear below the baseline, but it would be read the same way.

Ac signals can also be read directly from the oscilloscope. Switch your input coupling control to its ac position, then set a baseline just as you would for dc measurements. If you are unsure how to set the vertical sensitivity, start with a low sensitivity (a large VOLTS/DIV setting), then slowly increase the sensitivity (reduce the VOLTS/DIV setting) once your input signal is connected.

Keep in mind that ac voltage measurements on an oscilloscope will not match ac voltage readings on a multimeter. An oscilloscope displays instantaneous peak values for a waveform, while ac voltmeters measure in terms of rms (root mean square) values. To convert a peak voltage reading to rms, divide the peak reading by 1.414. Another limitation of multimeters is that they can only measure sinusoidal ac signals. Square, triangle, or other unusual waveforms will not be interpreted correctly by a multimeter.

When actually measuring an ac signal, it might be necessary to adjust the oscilloscope's trigger level control to obtain a stable (still) trace. As Fig. 4-18 illustrates, signal voltages can be measured directly from the display. For example, the sinusoidal waveform of Fig. 4-18 varies from −10 to +10 V. If the oscilloscope sensitivity was set to 5 VOLTS/DIV, signal peaks would occur 2 divisions above and 2 divisions below the baseline. An oscilloscope provides peak measurements, but an ac voltmeter would show the signal as peak/1.414 .

[10/1.414] or 7.07 Vrms

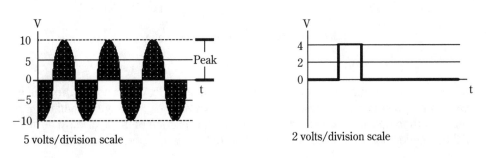

5 volts/division scale   2 volts/division scale

4-18 Reading ac voltage signals on an oscilloscope.

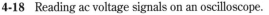

Time and frequency measurements

An oscilloscope is an ideal tool for measuring critical parameters such as pulse width, duty cycle, and frequency. The horizontal sensitivity control (TIME/DIV) comes into play with time and frequency measurements. Before making any measurements, you must first obtain a clear baseline as you would for voltage measurements. When a baseline is established and a signal is finally connected, adjust the TIME/DIV control to display one or two complete signal cycles.

Typical period measurements are illustrated in Fig. 4-19. With VOLTS/DIV set to 5 ms/DIV, the sinusoidal waveform shown repeats every 2 divisions. This represents a period of

[5 ms/DIV × 2 DIV] 10 ms

Because frequency is simply the reciprocal of time, it can be calculated by inverting the time value. A period of 10 ms would represent a frequency of

[1/10 ms] 100 Hz

This also works for square waves and regularly repeating non-sinusoidal waveforms. The square wave shown in Fig. 4-19 repeats every 4 divisions. At a TIME/DIV setting of 1 ms/DIV, its period would be 4 ms. This corresponds to a frequency of

[1/4 ms] 250 Hz

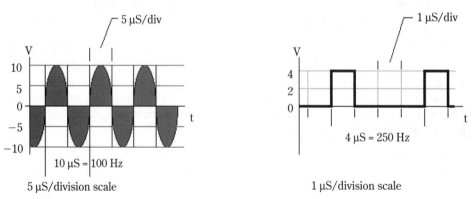

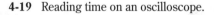

4-19   Reading time on an oscilloscope.

Instead of measuring the entire period of a pulse cycle, you can read the time between any two points of interest. For the square wave of Fig. 4-19, you could read the pulse width to be 1 ms. You could also read the low portion of the cycle as a duration of 3 ms (added together for its total signal period of 4 ms). A signal's *duty cycle* is simply the ratio of a signal's on time to its total period expressed as a percentage. For example, a square wave that is on for 2 ms and off for 2 ms would have a duty cycle of

[2 ms/(2 ms + 2 ms) × 100%] 50%

For an on time of 1 ms and an off time of 3 ms, its duty cycle would be

[1 ms/(1 ms + 3 ms) × 100%] 25%

and so on.

Specialized equipment

Multimeters, logic probes, and oscilloscopes are by no means the only test equipment that have uses in telephone repairs. These are merely general-purpose instruments that can aid you in specific electrical and electronic troubleshooting. There is also a myriad of specialized telephone test equipment that can be used in applications ranging from simple phone line checks, to economy telephone signal generator/testers, to feature-packed central office simulators. Figures 4-20, 4-21, and 4-22 show examples of such equipment.

On many occasions, it is desirable to test a telephone without actually connecting it to a live telephone line. Telephone test equipment can provide simulated

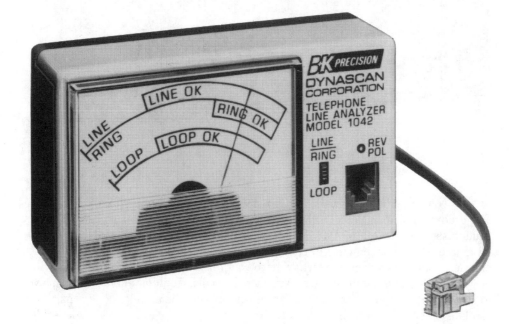

4-20 A telephone line analyzer. Courtesy of B+K Precision.

4-21   A telephone product tester. Courtesy of B+K Precision.

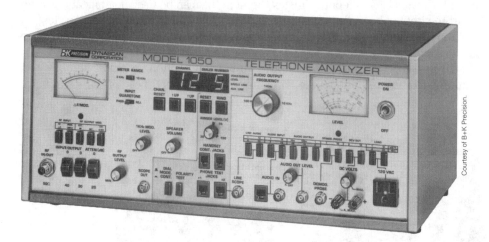

4-22   A telephone product analyzer.

battery voltage, ringing signals, and dial tone to the telephone you are repairing. Some equipment can detect voice and dial signal levels to ensure proper voice transmission volume and DTMF tone pair settings. Telephone test equipment eliminates the inconvenience of calling others to test the speech path and dialing characteristics of your phone or having to wait for others to call you.

For the beginner or one-time repair, such specialized equipment is too expensive to justify space on your workbench. For individuals practicing telephone repair as a sideline or second career, however, this type of equipment can prove indispensable. For the purposes of this book, coverage of specialized test equipment ends here. However, it is important to know that such equipment does exist, and that it is available commercially.

5
Wiring installation and repair

BEFORE YOU ACTUALLY BEGIN TO WORK WITH TELEPHONES (FIG. 5-1), IT HELPS
to understand the wiring in and around your home or apartment that connects your
telephone to the PSTN. This chapter will also introduce you to a selection of hand
tools you can use to install and make repairs to telephone wiring. It describes some
techniques for making and repairing line and handset cords, and tells you how to install a number of modern telephone jacks.

The network interface

Things used to be so simple. Once upon a time, there was a major corporation called
"the phone company". The phone company (American Telephone & Telegraph Company, or AT&T) used to handle every last detail of telephone wiring—from the central office, right down to the individual telephone jacks in your home. If you needed
to have wiring or a telephone repaired, the phone company would come right out and
assist you—free (service was a part of your monthly bill).

With government deregulation of the telecommunications industry and the
breakup of "the phone company" into smaller, regional corporations called Bell Operating Companies (BOCs or *baby bells*), your regional telephone company no longer
includes in-home wiring or service as part of their routine services. Today, your BOC
simply carries your telephone line from a telephone pole (or underground) outside,
and terminates the line in a specially designed connector called the network interface
as shown in Fig. 5-2. All the wiring in your home originates from this connector. The
BOCs will do home wiring work, but they charge a steep hourly fee. Most telephone
wiring for new homes is provided by independent electrical contractors, but there is
no reason why you cannot install and repair your own telephone wiring.

5-1 A dual-cassette answering machine. Courtesy of Radio Shack.

Older telephone installations (prior to the breakup of AT&T) did not always use a network interface connector. Instead, the telephone line from the pole might simply terminate at a set of screw terminals in your basement as shown in Fig. 5-3. Although the method of connection might be different, the key item to remember is that a telephone line enters your home at one point only (usually in the basement). From there, it is made available to the telephone jack(s) in your home. If you carry an extra telephone line (perhaps for a home business), the second line enters at the same point, but it terminates on its own network interface or another set of screw terminals.

As a general rule, the BOC is responsible for maintaining your service up to and including the point where the line enters your home (at the network interface). For wiring and repairs inside of your home, you are on your own.

Installations

Telephone wiring is a remarkably straightforward process. You will only need a few simple tools to do the job quickly and efficiently.

Tools and materials

In many cases, a medium-size regular-blade screwdriver and a pair of reliable wire cutters/strippers are enough. You might want to use an industrial-grade staple gun

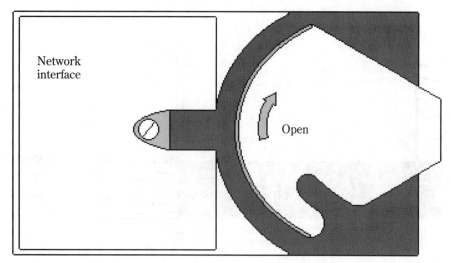

5-2 A network interface connector.

or nail-in staples to secure exposed wiring. For routing cables through walls or floors, use a power drill with a long drill bit.

If you are going to make modular telephone line or handset cables, you will need a modular telephone crimping tool, 4-wire telephone cable (flat stock), and a supply of appropriate modular crimp-on connectors. Telephones use two types of modular connectors: smaller, 4-position connectors for the handset, and slightly larger 6-position connectors for the line cord. Even though the line cord uses 6-position connectors, only the center 4 positions are used. Handset and line cord connectors are different, even though they might look identical at first glance. Many mail-order electronics houses and consumer electronics stores sell crimping tools and connectors.

For new installations, you will also need R-type modular receptacles. These are readily available in a selection of shapes and sizes depending on how and where you wish to mount them. As Fig. 5-4 shows, modular receptacles are categorized as *surface mount* or *flush mount*.

Telephone wire

Standard telephone wire is available from most mail-order or consumer electronic stores either by the foot, or in precut bulk rolls. The term *telephone wire* is rather misleading. It is actually a cable containing at least 4 (maybe 6 or 8) solid conductors arranged in a flat or round fashion—as shown in Fig. 5-5. Round cable is usually used to wire modular receptacles throughout a home, while flat cable is used to handle crimp-on modular connectors.

As you recall from chapter 1, the green and red wires of a telephone cable are traditionally used to mark the tip and ring leads respectively. These are the active, signal-carrying conductors in your home wiring. The cable's black wire is sometimes employed as a telephone ground, and the cable's yellow wire can serve as a

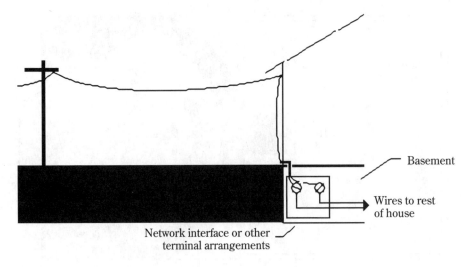

5-3   A typical phone line entrance.

signalling or control lead. The use of black and yellow wires in a telephone line is very inconsistent.

Making modular line cords

The *line cord* is the cable that connects the telephone to a receptacle. Such a line cord is usually terminated with modular connectors at both ends, although some older cords have spade lugs on the wires of one end.

You will need several items to construct a line cord: an appropriate length of flat, 4-conductor line cord; two, 6-position (not 4-position) modular connectors; an appropriate modular crimping tool which can accommodate a 6-position modular connector (many commercial modular crimping tools are designed to handle 4- or 6-position connectors); and a razor knife or suitable wire stripper for flat telephone cable.

Use your razor knife or wire stripper to cut away ³⁄₁₆ to ¼ inch of the cable's outer jacket as shown in Fig. 5-6. Do not strip away any of the insulation from individually-colored wires. Place this prepared cable into the open end of a modular line cord connector. Make sure that each individual wire is inserted properly under each corresponding electrical contact. Because line cords use 6-position modular connectors with 4-conductor cable, make sure to use the center 4 positions of the connector.

The outer jacket of the cable should be inserted far enough into the body of the connector so that the jacket is present under the cable pinch point. If too much of the outer jacket has been stripped away, trim off the individual wires until the cable fits correctly. When crimping takes place, the cable pinch point of the connector is crushed against the outer jacket—this applies pressure to hold the cable in place.

After the cable has been properly prepared and seated in the modular connector, insert the entire assembly into the appropriate orifice of your crimping tool. Squeeze the crimping tool completely. The instructions accompanying your particular crimping

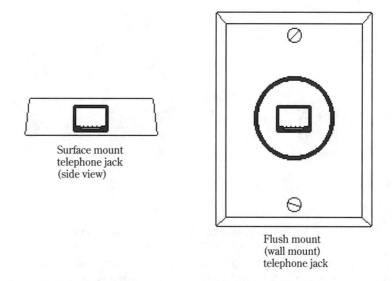

Surface mount
telephone jack
(side view)

Flush mount
(wall mount)
telephone jack

5-4   Comparison of two typical modular receptacles.

tool will be more specific—different tools have subtle differences in the way crimps are made. The process of crimping forces each electrical contact pin to penetrate its corresponding conductor and make contact with the copper wire inside. Once a crimp is made, it is permanent, and can only be removed by cutting off the connector.

Telephone line cords are typically *straight through* cables, as shown in Fig. 5-7. If you were to hold each modular connector side by side, it would seem that the wire order is reversed. In reality, however, the cable is terminated without any reversals or flips.

Making handset cords

The *handset cord* is the cable that connects the main body of a telephone to a removable hand-held assembly. As a minimum, a handset assembly must house a transmitter and receiver. Some telephones house much more in a handset—such as amplifiers and a dial pad. (Of course, cordless telephones do not use handset cords.) Some older telephones use handset cords terminated with spade lugs at both ends, but newer designs use the modular handset cords almost exclusively. For the purpose of this book, you can assume a modular connector at both ends of the cable. You will probably never have to build a modular handset cord from scratch, but you might find yourself repairing a occasional intermittent or open cord.

Before you take any action with a handset cord, make note of the wire color sequence at the suspect connector. This way, you will be certain to install a new connector in exactly the same fashion as the old one. Cut off the suspect connector and discard it (clip the handset cord as close to the connector as possible). Use a razor knife or wire stripper to cut away ³⁄₁₆ to ¼ inch of the cable's outer jacket (similar to what is shown in Fig. 5-6). Do not strip any insulation from the individually-colored wires. Place the prepared cable into the open end of a 4-position modular handset

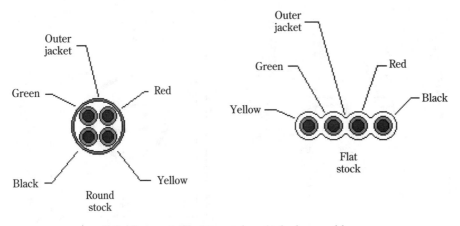

5-5   Comparison of two styles of telephone cable.

connector. When you are performing a repair, make sure that the individually-colored wires are inserted in the same order as the original connector, and are seated properly under each corresponding electrical crimp contact. The outer jacket should be inserted far enough into the connector's body so that jacket material is available under the connector's cable pinch point. If too much of the outer jacket has been stripped away, trim the individual wires until the cable fits correctly. When crimping takes place, the cable pinch point is crushed to apply pressure to the outer jacket— this pressure holds the cable in place.

After the handset cord is prepared and correctly inserted into its modular connector, load the entire assembly into the appropriate orifice of your crimping tool. Squeeze the crimping tool completely to perform a thorough crimp. The instructions accompanying your particular crimping tool will be more specific—various tools have subtle differences in the way crimps are made. The process of crimping forces each electrical contact to pierce the insulation of each individual wire and make contact with the conductor inside.

Use caution when constructing a handset cord. Some cords flip certain wires between both connectors. Make a careful note of the wire colors and configuration before cutting off the old connector.

Installing a modular receptacle

Receptacles (or *phone jacks*) can be installed either as part of a new construction, or to add a new receptacle to an existing home or apartment. You will need a receptacle (surface or flush mount), one or two appropriate mounting screws (usually enclosed with the receptacle), a medium-sized regular-blade screwdriver, and a length of round, 4-conductor telephone cable. An exposed view of a standard, surface mount receptacle is shown in Fig. 5-8. Notice that the receptacle is pre-wired to its connector leads with black, green, red, and yellow wires. If you look closely at the receptacle, you will discover that it is actually built to handle six wires (remember from the preceding section that a telephone line cord uses 6-position connectors), but only 4 wires are used. The additional two terminals (blue and white) are not used.

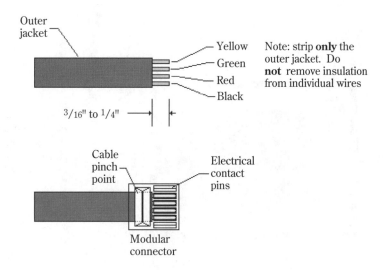

5-6   Preparing a modular connector.

Select a location for the receptacle—preferably on a low, wooden baseboard or molding. If you want to mount the receptacle in a sheetrock or plaster surface, you might need to use plastic screw anchors to support the insertion of screws, or find some other means of attaching the receptacle to the desired surface. Some receptacles require one screw to secure it. Other receptacles (such as flush mount receptacles) require two screws.

Once the new receptacle is secure, run the 4-conductor telephone cable from its source to the new receptacle. Make sure to leave 5 to 8 inches of extra cable as a *service loop*. Extra cable ensures that you will have enough wire available if you ever need to cut down the wiring during a repair. After the individual wires are attached, you can coil the service loop inside the receptacle's cover. For new construction, you might be able to stuff the extra wire length into the wall space. If this installation is a post-construction addition, you should dress the exposed cable using a good-quality, industrial-grade staple gun. Be careful to do a neat job.

Once the cable is run and dressed, connect the individual wires at the source and new receptacle as shown in Fig. 5-9. Use a pair of wire strippers to remove about 4 inches of the outer jacket from each end of the cable. Then, remove about ½ to ¾ inch of insulation from each of the individual conductors. Loosen each screw terminal at the receptacle, then wrap the copper portion of each wire in a clockwise direction around its correspondingly-colored screw terminal. For example, attach the cable's red wire to the receptacle's red terminal, etc.

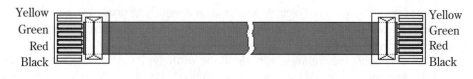

5-7   A straight-through cable assembly.

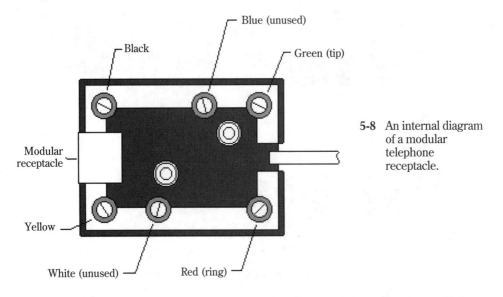

Black

Blue (unused)

Green (tip)

Modular
receptacle

5-8   An internal diagram
of a modular
telephone
receptacle.

Yellow

White (unused)

Red (ring)

Finally, tighten each screw terminal in the clockwise direction. After all four wires have been secured, coil the remaining service loop carefully inside the receptacle's cover, then replace the cover. You can also push the remaining service loop back into an empty wall space beneath the receptacle.

The process of installing a wall (flush mount) receptacle is exactly the same as that of the surface mount receptacle. However, you can expect to encounter added complexity if you decide to run telephone cable inside walls or other structures. It is up to you just how much effort you want to put into making your installation esthetically pleasing.

Repairing home wiring

Home telephone wiring rarely fails under normal circumstances. Once the wiring is installed and working properly, there is little that can go wrong. This concept of reliability is the deceptive aspect of home wiring—it fails so rarely that it is often overlooked as a potential source of trouble. As a result, the telephone itself might be wrongly assumed to be at fault even though a wiring problem is actually to blame. This section of the chapter shows you how to determine whether the telephone or wiring is at fault, and how to track down wiring problems.

Telephone wiring can fail for several reasons—metal corrosion due to moisture, busy rodents at work in your basement or house structure, and curious infants pulling on exposed wiring or receptacles are only some of the natural disasters that can plague telephone wiring. To carry out the troubleshooting procedures in this section, you will need a multimeter, and a *butt adaptor,* which you can make very easily (as shown in Fig. 5-10).

There are many places along your telephone wiring you might want to test—but a modular telephone can only connect to corresponding modular receptacles. To overcome this, you can wire regular alligator leads into an extra modular receptacle.

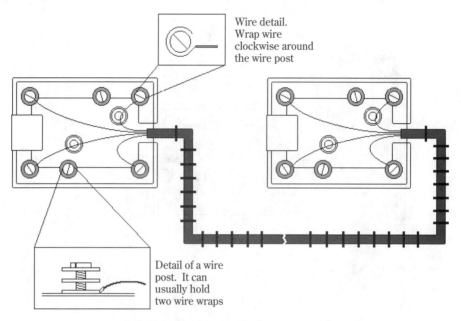

Wire detail. Wrap wire clockwise around the wire post

Detail of a wire post. It can usually hold two wire wraps

5-9   Installing a new telephone receptacle.

That way, you need only plug the telephone into your makeshift connector, then simply clip the alligator test leads across tip and ring at any points you wish to measure. To make such a butt adaptor easier to visualize, use a green alligator clip to represent tip, and a red alligator clip for ring.

It is strongly recommended that you use a very inexpensive telephone during your wiring troubleshooting. Because you will be moving the telephone around quite a bit, the test telephone could easily be dropped or abused several times during your testing. Simple, inexpensive telephones can often resist accidental abuse better than more sophisticated and delicate telephones.

Isolating the telephone

When you encounter telephone problems, you must first determine whether the telephone, your home wiring, or the BOC's wiring is at fault. Remember, your telephone itself is merely a small portion of a massive, worldwide network. If the network fails, so will your telephone.

Before you begin any sort of a repair, you must isolate the trouble to either your telephone, or some portion of the telephone network (either inside or outside of your home). The first method of isolation is to test the telephone in a known good working receptacle. Perhaps a neighbor might be willing to let you try your telephone in one of their telephone jacks. If you have a second telephone in your home, and that unit is working normally, you can try a suspect telephone there as well. If the telephone continues to malfunction in a known good receptacle (known good because another telephone works there just fine), then the suspect telephone itself is probably defective. You can then begin to apply specific troubleshooting procedures to repair the defective telephone.

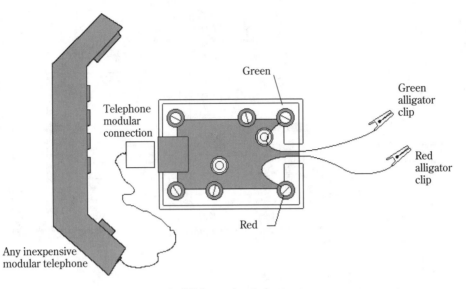

Green

Green
alligator
clip

Telephone
modular
connection

Red
alligator
clip

Red

Any inexpensive
modular telephone

5-10 Wiring a simple butt set.

If, however, the suspect telephone works properly in a known good receptacle, the problem is most likely at some point in the network. At this point, you will have to determine whether the trouble is in your home wiring, or somewhere in your BOC's wiring. Take your working telephone and plug it into the network interface connector, then try your telephone. When a telephone works as expected in the network interface, your home's wiring is probably defective. Questionable telephone operation at the network interface suggests a problem with the BOC's system. If your home is not fitted with a modular network interface, you can use the butt adaptor setup as shown in Fig. 5-10 to tap into the main junction—the point where the telephone line first enters your home.

It is very important that you include the telephone's original line cord as you move a telephone from place to place. The stress and abuse sustained by most line cords can easily result in intermittent or open modular connectors. If a telephone seems to be defective, try a new line cord before starting a telephone repair.

This troubleshooting process is illustrated graphically in the chart of Fig. 5-11. Using this methodology, you can determine whether it is the telephone, home wiring, or outside wiring that is faulty. If you determine that the outside wiring is faulty, place a call to your local BOC—they are responsible for servicing the network outside of your home. If you think your internal home wiring might be defective, refer to the next section for more specific troubleshooting procedures.

Troubleshooting home wiring

In principle, telephone wiring is remarkably straightforward (Fig. 5-12). The signal-carrying wires are provided from your local CO, and routed into your home to a single main junction or network interface connector. From there, tip and ring are provided to one or more of the modular receptacles located around your home. The amount and complexity of the wiring depends on how large your home is, and how

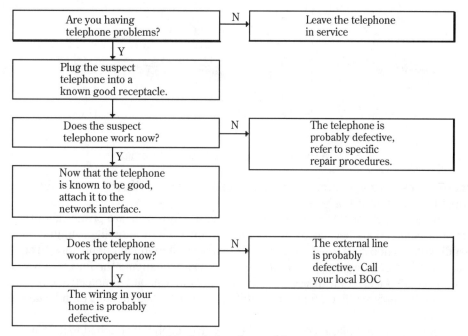

5-11 Isolating the trouble source.

many receptacles are installed. Electrically speaking, every telephone is installed in parallel across the signal-carrying wires.

Two other wires—black and yellow—are often included with the green and red wires of tip and ring. However, these extra wires are not used consistently, so you should concern yourself only with tip and ring. Keep in mind that there is no limit to the number of telephone receptacles that can be installed in a home, but only one telephone can be in use at any one time.

Unfortunately, home wiring is rarely as simple as Fig. 5-12 shows. In actual practice, a typical installation is somewhat more involved, such as the wiring shown in Fig. 5-13. This is, by no means, the only possible set of connections. The layout of your own wiring is probably much different, but the principles will always be the same. Figure 5-13 will serve as the model for the following troubleshooting procedures.

Although each of the telephones shown in Fig. 5-13 are connected in parallel across tip and ring (just like the telephones of Fig. 5-12), you can see that the wiring in the layout of Fig. 5-13 is much more involved. Multiple parallel branches are used to route the second floor telephones into the network. As you can see, breaks at various points in the wiring can adversely affect different sets of telephones. Wiring problems are generally the result of faulty electrical connections either in the cables, line cords, or connectors.

Symptom 1   You hear random or intermittent static on one or more telephones. The telephones themselves are assumed to be working properly. Static in wiring is almost always caused by a poor connection which has not yet failed completely. It can often be stimulated by wiggling an adjacent line cord or telephone cable, or by tapping on a nearby receptacle. The degree of static depends on the condition of the particular connection at fault. Static can vary from occasional, barely perceivable

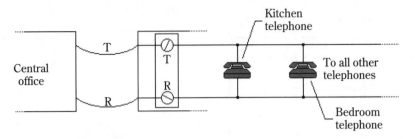

5-12 A simplified diagram of telephone wiring.

interruptions in reception or transmission, to a constant, irritating crackle that renders the telephone virtually useless.

Check your suspect telephone's line cord first. Line cords are prime candidates for poor connections because of the fatigue produced by the pulling, tugging, and bending that a line cord must undergo. Remember that modular connectors on each end of the line cord are only crimped into place. Try exchanging the line cord with a new or rebuilt one. If static disappears, the line cord was at fault. A defective line cord can often be repaired by cutting off the defective modular connector, then installing a new one.

Inspect your modular receptacle and the adjacent cabling attached to it. Remove the outer cover of the receptacle and gently wiggle each of the colored wires attached to screw terminals. Receptacles and their associated wiring are usually stationary, so it is unusual for connections to fail—unless they were not made properly in the first place. Tighten any screw terminals that appear to be loose. Reseat the wires on each terminal if necessary. If you inspect the modular connector on the receptacle, you will see four metal prongs, as shown in Fig. 5-14. If any of these prongs are bent or loose, they will not make good contact with the connectors of a line cord. Replace any receptacle that has defective electrical contacts.

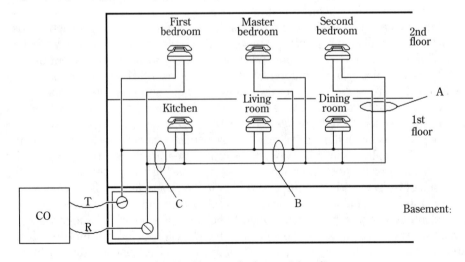

5-13 A typical home telephone wiring diagram.

If any electrical contacts are loose or bent, a faulty or defective connection may result

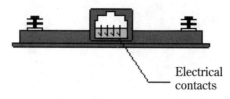

Electrical contacts

5-14 The connector-view of a standard modular receptacle.

Symptom 2   One or more telephones are completely dead—there is no sidetone or dial tone. The telephones themselves are assumed to be working properly.

As you saw in Fig. 5-13, each telephone is wired in parallel across the tip and ring provided from a central office. An open circuit in the wiring due to a broken connection is usually responsible for a dead telephone. However, the location of the wiring problem can have a profound impact on just how many telephones are affected.

For example, suppose a wiring failure occurred at point A in Fig. 5-13. Such a failure would disable the telephone in the second bedroom. However, suppose the same wiring failure occurs at point B. A failure at that location would disable telephones in the master bedroom, dining room, and second bedroom. A wiring failure at point C would render all telephones inoperative except for the one in the first bedroom. Knowing how your home is wired will greatly simplify the task of troubleshooting.

Unfortunately, it is not always possible to know the telephone wiring in your home. Start at your defective telephone and follow the wiring until you find a telephone that works (or until you reach the network interface). Measure the dc voltage across the tip and ring (green and red respectively) wires of the defective receptacle with all telephones in the home onhook (idle).

If you read –48 Vdc, the wiring to the receptacle is probably good, but the modular line cord or its connector might be defective. Try a new line cord first, then try installing a new modular receptacle. If –48 Vdc is absent from the suspect location, there is a wiring break between the working and defective receptacle. You can attempt to locate and re-splice the faulty wire(s)—or run a length of new wire between the receptacles to jumper out any suspect wiring.

6
Classical telephones

NOW THAT YOU HAVE SEEN SOME FUNDAMENTAL ASPECTS OF TELEPHONE technology, learned how to identify and read basic electronic components, covered a variety of soldering and printed circuit service guidelines, and read about the application of simple test equipment, you can begin to use that information to troubleshoot and repair telephones and answering machines.

The best place to start is by understanding the components, operations, and troubleshooting procedures of classical telephones (Fig. 6-1). Telecommunications technology has evolved so quickly with the widespread use of integrated circuits, that it is difficult to precisely define a classical telephone. The telephones and answering machines designed and manufactured today are much more powerful and far more versatile than those made just a few short years ago. However, we have to start somewhere, so a definition must be drawn. For the purposes of this book, classical telephones are electromechanical devices that use solid-state components up to and including transistors—but have no integrated circuits. Telephones that contain at least one integrated circuit are called *electronic telephones*—they will be discussed in their various forms throughout the rest of this book.

The classical approach

A classical telephone can be broken down into roughly six functional areas, as shown in the block diagram of Fig. 6-2: the hook switch, the ringer, the dialer, the receiver, the transmitter, and the speech circuit. Each of these sections is vitally important to the operation of every telephone. Electronic, cordless, and cellular telephones also require these basic areas, but their circuitry is much more advanced, and other functions are

Courtesy of Radio Shack.

6-1 A classical telephone.

needed to complete their operation. This part of the chapter describes and details the circuitry used in classical telephones.

The block diagram of a rotary pulse-dialing telephone in Fig. 6-2 is a simplified version of the schematic shown in Fig. 6-3. While this circuit might appear somewhat confusing at first, it is actually remarkably straightforward.

Hook switch

When you inspect the circuit closely, you will see that it consists of two distinct switch assemblies: the hook switch and the contact sets used by the rotary dialer. As you read in chapter 2, the hook switch is responsible for connecting and disconnecting a telephone from the PSTN. With classical telephones, a hook switch is actuated by the weight of a handset. If the handset is lifted, hook switch contacts 1 and 2, as well as contacts 3 and 4, close to energize the speech circuit (also known as the *network*). Contacts 5 and 6 open when the handset is lifted. This allows signals in the speech circuit to be heard at the receiver.

Rotary dialer

Notice that the rotary dial uses two sets of contacts. The contacts marked *mute* are placed across the receiver. Whenever the dial plate is moved away from its rest posi-

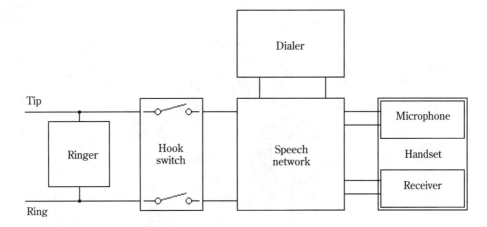

6-2 A classical telephone.

tion, the mute contacts close to short-circuit any signals that might otherwise reach the receiver. This shorting action prevents loud, annoying click sounds from being heard while the dialer is in motion. The second set of contacts (marked *pulse*) open and close while the dial plate rotates. When the pulse contacts open, the speech circuit is interrupted, so no loop current can flow. When the pulse contacts close, the circuit draws normal loop current from the local central office. When at rest, pulse contacts are normally closed to complete the speech circuit.

When the dial plate is moved from its rest position, the movement applies tension to a spring. You can feel this tension as you advance the dial plate clockwise. After the dial plate is released, spring tension provides the mechanical energy to rotate the dial plate counterclockwise toward its rest position. A mechanical governor is added to ensure that the dial moves at a constant rate—no matter how far the plate is advanced.

A lobed cam is included in the dialer's mechanical assembly to open and close the pulse contacts. Because the dial turns at a constant speed, the cam assembly opens and closes the pulse contacts at a fixed rate (specified in pulses per second, or pps). Rotary dialing units operate at roughly 10 pps, so the duration of one opening and closing cycle is about 0.1 second (or 100 milliseconds). For each cycle, the pulse contacts are open for a fixed amount of time (called the *break time*), and closed for a fixed amount of time (called the *make time*). Make and break times are defined by the shape of the particular cam in use. In the United States, contacts are usually broken for 60 milliseconds, and made for 40 milliseconds:

60 ms + 40 ms = 100 ms

Such a cycle is commonly referred to as a 60/40 ratio. In other countries, the ratio may be 67/33.

One problem with rotary dialers is that they can produce huge voltage spikes each time opening pulse contacts interrupt the flow of current. These high-voltage spikes would often cause the ringer to sound softly—rather like a tinkle. Voltage spikes would also be heard as loud clicks in the receiver portion of the handset. To

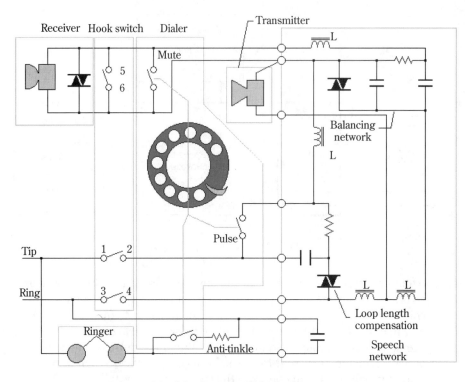

6-3 A basic rotary dial telephone.

prevent tinkle and save the caller from certain annoyance, an anti-tinkle scheme was adopted. When the dial plate is moved from its rest position, the mute contacts close to short-circuit the receiver. Another switch closes to connect a low-value resistor, called an *anti-tinkle resistor*, across the ringer coil. Both of these switches remain closed while the dial plate is in motion. The ringer capacitor acts to filter any high-voltage transients that pulse contacts generate.

Tone dialer

For many years, switching at your local central office was performed by electrome-chanical equipment. Such switching equipment operated directly from the dial pulses received from rotary dialing units. As years passed, however, people sought faster, more convenient ways to specify selected digits.

Telephone system designers came up with a dialing method that uses audible tones instead of pulses. Each desired digit (0 through 9) is represented by a set of two unique tones. Because two audible tones are used, the method became known as *dual-tone multifrequency* (DTMF) dialing. You are probably familiar with the trade name for this method: "TouchTone." There are several distinct advantages to using dual tones. First, a precise combination of two tones is much harder to du-plicate accidentally than a single tone, so it is virtually impossible for a person to cause false dialing through normal operation. Second, 12 or 16 unique frequency combinations can be generated using only seven tone oscillators, so less overhead circuitry is required than for a dialer generating tones at 12 (or 16) separate frequencies.

Figure 6-4 illustrates the standard key layout and frequency assignments for a regular 12-key DTMF dial pad. As an example, when the 5 key is pressed, the dialer produces a mix of 770-Hz and 1336-Hz tones. This mix of frequencies is then filtered and interpreted by the local central office. Each of the DTMF dialer's frequencies are internationally standardized. In North America, each frequency must be kept within a tight tolerance of ±1.5%.

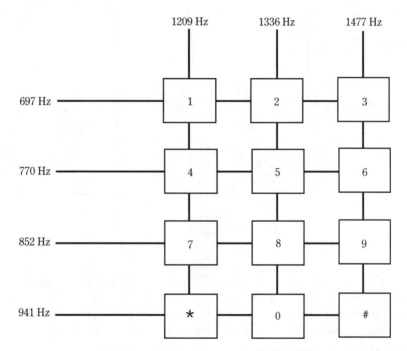

6-4 A standardized layout of a DTMF dial pad.

Each tone is generated by an oscillator circuit, as shown in the diagram of Fig. 6-5. The diagram represents a discrete oscillator configuration using LC (inductor-capacitor) combinations to create resonant circuits. When no keys are pressed, all dial pad switch contacts are open. This effectively disconnects the resonating capacitor from any portion of the inductors, so no resonant circuits are created. When a key is pressed, a mechanical linkage closes both a *row* and a *column* contact, creating two resonant circuits. The transistor is connected into the circuit by switch C whenever any key is pressed. This sustains and mixes the oscillations of both resonant circuits. The DTMF signal is then connected to the telephone line. When the desired key is released, both oscillators are shut down again. Many integrated circuits provide DTMF signals in today's electronic telephones—but discrete LC oscillators are used exclusively for DTMF signalling in classical telephones.

As far as the telephone's user is concerned, DTMF signalling is much faster than rotary dialing. For example, any DTMF digit can be sent and interpreted by

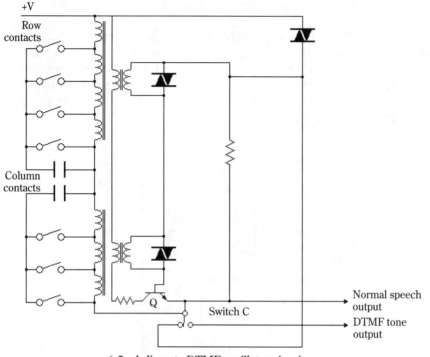

6-5   A discrete DTMF oscillator circuit.

a central office in about 100 milliseconds. This time includes the actual time to interpret the digit, as well as a certain amount of time required between digits (known as *inter-digit delay*). Compare a brief 100 ms per digit to the 100 ms required for each rotary pulse—dialing a 2 would require 200 ms, a 5 would require 500 ms, and so on.

Like most pieces of electronic equipment, DTMF dialers must receive the appropriate levels of voltage and current in order to operate properly. Proper operation guarantees that every generated tone has the required amplitude and low-distortion characteristics to be accurately detected by a central office. The problem of circuit power is further complicated by the fact that classical DTMF circuits must be powered from the telephone line itself. The voltage and current available at your particular telephone is strongly related to loop length (the telephone's physical distance from its central office). So it is critical that the DTMF circuit be able to function within a wide range of voltage and current levels.

Long telephone loops cause large losses at your telephone—loop voltage and current at your telephone might be well below their ideal values. DTMF circuits are designed to produce acceptable tones with as little as 3 Vdc across tip and ring. Short loops often present the opposite problem—a telephone close to its central office facilities can receive an excess of voltage and current from the telephone line. Luckily, classical DTMF dialers are typically immune to such over-voltage conditions, and electronic dialer circuits typically incorporate some form of voltage regulation to keep IC power at a constant level.

Ringer

The electromechanical ringer assembly is placed directly across tip and ring. Notice that a capacitor is placed in series with the ringer. The capacitor serves to block dc from energizing the ringer. Resulting current drawn by an unblocked ringer would fool the local central office into thinking that the telephone is offhook—even if the telephone is idle.

Speech circuit

The key to any classical telephone is its speech circuit. A speech circuit is responsible for a variety of functions. It must couple an electromagnetic receiver, carbon microphone, and dialer (rotary or DTMF) into the two-wire (tip and ring) telephone line, and provide adequate sidetone between the microphone and receiver. It must automatically compensate for variations in loop current due to line length. Finally, the speech circuit must support protective functions such as anti-tinkle and speech muting. Let's look at each of these functions individually.

Both the receiver and microphone are two-wire devices. This means that four wires must be translated into a two-wire circuit for transmission over the telephone line, while still allowing full-duplex operation (speaking and listening simultaneously). The translation from four to two wires is achieved in classical telephones using an induction coil, also known as a *hybrid*.

A typical hybrid is illustrated in Fig. 6-6. In reality, it is not just one, but two interconnected transformers wound on the same physical structure (or *core*). As with all transformers, a hybrid uses the principle of electromagnetic coupling to combine or separate transmitted and received signals. As an example, suppose a voice signal was applied to the transmit input. The changing voice signals in windings 1 and 2 produce varying magnetic fields. In turn, the magnetic fields induce corresponding secondary voltages across windings 3 and 4. The ratio of these *input voltages* to the corresponding *output voltages* depends on the number of turns of wire in the primary coil(s) versus the number of turns of wire in the secondary coil(s). Turns ratios are important in impedance matching between two coupled circuits.

The secondary signal at winding 3 is processed through a passive balancing network, then connected to winding 5. The secondary signal coupled to winding 4 becomes the transmitted signal sent along to the central office. Any signals that are received from the central office become the primary signal on winding 6. The magnetic field generated in winding 6 couples to winding 8. The resulting signal across winding 8 becomes the main signal heard in your receiver. Remember, however, that a small portion of your transmitted signal is available at winding 5. Just how much signal will depend on the adjustments of your specific balance network. This sample of the transmitted signal is coupled to winding 7. In effect, this coupling process adds a small portion of your voice to the signal heard in your receiver—sidetone is therefore created due to the intentionally slight imbalance in the speech circuit's balancing network.

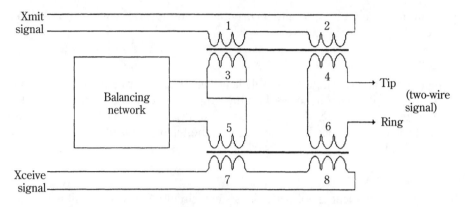

6-6　A typical hybrid transformer configuration.

By hearing your own voice, you can often judge how loudly to speak so that a caller can hear you clearly. If there is too little sidetone, you will find yourself yelling into the telephone. If there is too much sidetone, you will tend to whisper. A varistor in the balancing network automatically adjusts the sidetone level for variations in loop current due to line length. As you might have guessed, each of the coils marked L in Fig. 6-3 are part of a single hybrid transformer.

Classical construction

This part of the chapter deals with the way in which a classical telephone's components are assembled into a working unit. Once you are familiar with classical telephone layout and assembly techniques, troubleshooting will be much easier.

There are two general approaches to classical telephone construction: desk style and wall style. Both styles contain essentially the same major components—the difference is merely in the outer housing and chassis construction. A desk-style telephone is shown in Fig. 6-7. The telephone contains all of the key assemblies needed to function: a hook switch, a ringer, a speech network, a dial unit (either rotary or pulse), and a handset assembly containing a microphone and receiver. The chassis and outer housing are designed to sit on any flat surface. A handset rests on top of the housing. A wall-style telephone contains the same basic assemblies, but the housing and chassis construction are designed to be mounted on a flat vertical surface. The handset hangs down from its cradle when a wall-style telephone is onhook.

All of the assemblies shown in Fig. 6-7 must be wired together before they will operate. Classical telephones typically use wires terminated with rugged spade lugs to form connections. A standard wiring diagram for a 2500-type (a classical DTMF) telephone is illustrated in Fig. 6-8. Figure 6-8 also lists a table of common wire colors. The wiring of your particular telephone might not be exactly the same, but it is very similar. A classical telephone with a rotary dialer is often referred to as a 500-type telephone set.

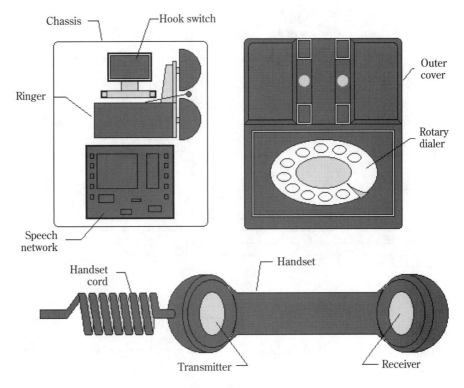

6-7　The breakdown of a classical desk telephone.

Troubleshooting classical telephones

Classical telephones are usually straightforward devices to troubleshoot. Their simple circuitry and interconnections can often be repaired in a matter of minutes. You might want to have your multimeter on hand to check loop voltage or wiring continuity at various points in the telephone. Before attempting to measure continuity, be certain that the telephone is unplugged. Disassembly can almost always be accomplished with little more than a regular or Phillips screwdriver. Remember to make careful notes of any wires you must remove during your repair. This will guarantee that you can reconnect the assemblies properly after the repair is made.

Symptom 1　The telephone is completely dead. No sidetone or CO dial tone is audible in the receiver. At first, it might not be clear whether the trouble lies in the telephone itself, in the house wiring, or in the line to the central office. Before you start disassembling the telephone, it is a good idea to confirm just where the problem actually is. One excellent method is to try a known operating telephone in the receptacle that the suspect telephone now occupies. If the known good telephone works properly, the suspect telephone is probably defective, and you can proceed with a repair. However, if a known good telephone also malfunctions, the problem might exist in your home wiring, or somewhere in your local

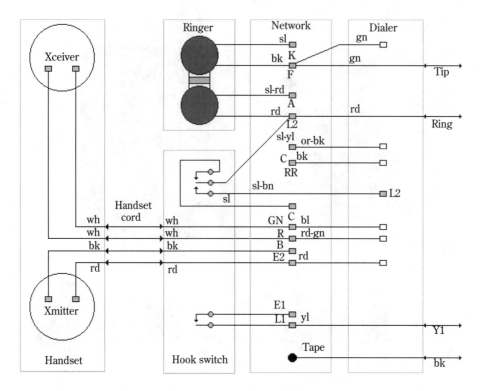

6-8   A wiring diagram of a typical 2500-type telephone set.

BOC's wiring. Refer to the procedures of chapter 5 for troubleshooting and repairing home wiring.

Be sure to check the telephone's line cord, especially if the line cord is modular. Crimped connectors that are yanked, twisted, and bent over a long period of time have a knack of becoming intermittent or open entirely.

After you are certain that the suspect telephone is indeed at fault (and not its line cord), you can begin the actual repair procedure. Remove the outer housing from the chassis. Once the outer housing is removed, the telephone's internal assemblies will be clearly exposed. Set your multimeter to a high dc voltage scale (perhaps 200 Vdc) and connect the multimeter across the tip and ring leads entering the telephone. Using Fig. 6-8 as a general guide (or drawing your own wiring layout), you can connect your multimeter at points F and L2 on the network PC board. When the telephone is onhook (idle), you should measure –48 Vdc (or +48 Vdc depending on your meter's lead orientation) across tip and ring. When the handset is lifted offhook (active), the telephone should draw loop current, and loop voltage should drop to about –6.5 Vdc (or +6.5 Vdc depending on your meter's lead orientation). The exact voltage values depend on the individual loop length—long loops yield lower voltages, while short loops tend to exhibit higher voltages.

If you witness the type of onhook/offhook response described above, then you know the telephone is active—it does draw loop current when offhook. When the telephone draws loop current as expected, the trouble usually lies in the handset cord. The handset cord, either modular or hard-wired, is subjected to a tremendous amount of stretching, bending, and flexing. There might be an open circuit at a connector, or in one of the wires. Try a new handset cord. If a new handset cord does not correct the problem, try a different electromechanical receiver. The receiver element might have become disconnected or failed internally.

If the handset cord and receiver appear to be intact, remove the speech network from its mountings and examine it carefully for any signs of cracks, missing, loose, or broken spade lug terminals, or other printed circuit damage. A fault in the network can cut off all audio signals from the handset. Inspect each lead of the hybrid. If the telephone has been abused or dropped, one or more of these leads could have broken or pulled away from the PC board. You can make repairs to most PC boards using the procedures presented in chapter 3.

If the telephone fails to draw loop current in its offhook state, then you know the telephone is not connecting to the central office. Check the connections between the telephone's line cord and the hook switch to ensure that there is continuity. Remove any protective coverings from the hook switch assembly, and carefully inspect each of the contact sets. Make sure that each contact set opens and closes cleanly as you paddle the hook switch on and off by hand. Clean the hook switch contacts with a clean cotton swab lightly dipped in alcohol or high-quality contact cleaner. Do not attempt to adjust the tension of any contact unless it is absolutely necessary. Finally, check the wiring from the hook switch to the network PC board.

Symptom 2   Ring is low (weak) or absent. As you read in chapter 2, electromechanical ringers are little more than electromagnets that drive a metal clapper. When a clapper swings, it strikes a metal gong to produce an audible sound. The pitch and tone of the resulting sound are directly related to the size and shape of the gong. The amplitude of a ring is dependent on the strength of the clapper strikes. Most electromechanical ringer designs incorporate a type of cam adjustment that affects the range of a clapper's swing. By limiting the clapper's range, you can reduce the force delivered to each gong, and thereby reduce the amplitude of a ringing signal. Before you open your telephone, check the chassis for a volume adjustment. Try a louder setting if possible.

Ringer solenoids are notoriously reliable devices—they are only energized occasionally, and then for short periods. There is not enough time for heat to accumulate and damage the coil. However, old age can sometimes cause the insulating enamel on solenoid wire to break down, and some older solenoids do develop short circuits between their various windings. As windings short-circuit, less magnetic force is generated to drive the clapper. Weakened solenoid forces result in weak (or absent) ringing.

Another possible problem is excessive coil resistance. A winding or internal connection might be opening in the ringer. The dc blocking capacitor can also fail. High-resistance circuits limit the current entering the solenoid. If there is not enough current to fully energize the coil, it cannot develop enough force to move the clapper properly.

Disconnect the telephone from its line cord, then use your multimeter to measure the resistance of the ringer solenoid as shown in Fig. 6-9. If there is more than one solenoid in the ringer assembly, measure both solenoids individually. A ringer coil can have a resistance of 5 to 50 Ω or more. The solenoid's resistance value is sometimes marked right on its body. If resistance level is extremely large or extremely small, the solenoid is probably defective. Try replacing the suspect ringer. Remember: this does not have to be an exact replacement part. The ringer from just about any older-model classical telephone should do just fine. Almost all electromechanical ringers are directly compatible with one another—after all, every ringer must work with the same central office ringing signal. The worst thing that can happen is that your telephone's ring might sound different.

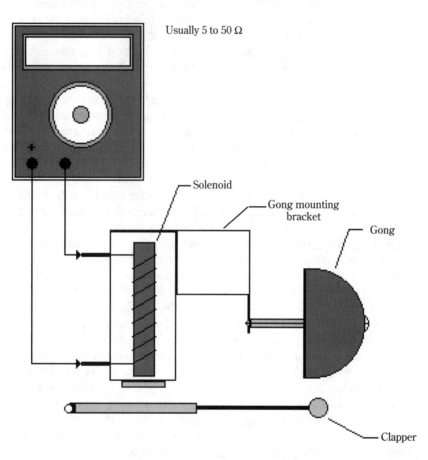

Usually 5 to 50 Ω

Solenoid

Gong mounting bracket

Gong

Clapper

6-9   Measuring resistance across a ringer solenoid.

A word of caution. Be certain to disconnect the line cord from your telephone before replacing the ringer. Ring signals from central offices in the United States often exceed 90 Vac. That is enough potential to deliver an uncomfortable, but almost never harmful, jolt should you be handling the ringer wiring and receive an incoming call.

If you are faced with no ring at all, measure the resistance across the ringer solenoid(s). If you detect infinite resistance across a solenoid coil, the coil has become an open circuit. An open circuit prevents any current from flowing, so no magnetic forces are generated to operate the clapper. Replace the ringer assembly.

On rare occasions, you might find a problem with the dc blocking capacitor in series with the ringer. You can use your multimeter to measure the capacitor's value as outlined in chapter 4. A failing (but not fully open-circuited) capacitor can result in a weak ringing signal. Replace the dc blocking capacitor, or replace the entire network PC board.

Symptom 3 Noise is present in the transmit or receive audio. Audible static on a telephone line can be maddening. Static can occur in your transmission, or your reception, or both. Before you assume that the telephone itself is at fault, test a known good working telephone in that location. If static symptoms persist on a known good telephone, the trouble could lie in your house, in the outside wiring to the local central office, or at any point between you and the calling telephone. If static symptoms disappear when using a known good telephone, you have probably isolated the trouble to the suspect telephone. You can then unplug the line cord and begin a repair. Remember to try exchanging the line cord of a suspect telephone. Line cords that are pulled, twisted, or bent regularly can develop intermittent connections that cause static symptoms.

Handset cords are also a regular source of static. The heavy strain most line cords are subjected to can eventually lead to fractures in the wiring or faulty connectors. If static seems to occur as you move the handset or cord, try replacing the handset cord. Modular cords are extremely simple to replace, but hard-wired handset cords require a bit more work—both the handset and telephone must be disassembled to replace hard-wired handset cords.

Closely inspect the speech network PC board for any signs of hairline cracks or failing solder connections. PC board damage can occur if the telephone has been dropped, or has been subjected to sudden, severe mechanical shocks. Classical speech networks are very simple circuits, so PC board repairs are usually straightforward. A low-power magnifying glass can help you to observe the PC board more thoroughly. Resolder any questionable solder points. Be sure to wear appropriate eye protection whenever you perform soldering operations. Chapter 3 outlines a variety of PC board repair procedures.

Another potential noise source is the telephone's hook switch. The telephone line is connected to the telephone through a set of hook switch contacts—any loose or dirty contacts can cause random interruptions in the circuit. Clean each hook switch contact carefully using a clean cotton swab dipped lightly in alcohol or high-quality contact cleaner. Do not adjust the tension or position of hook switch contacts unless it is absolutely necessary. Make sure that each wire from the hook switch assembly is securely attached to its appropriate place on the network PC board.

Symptom 4 Rotary dial problems. Rotary dial will not break dial tone once dialing pulses are sent. A rotary dial mechanism is truly a marvel of mechanical engineering—rugged, and elegant in its simplicity. However, mechanical parts in motion do wear out over time. Each part in a rotary dial is intended to open and close a set of electrical contacts for fixed periods at a constant rate. Typical rotary dialing specifications in the United States use a 60 ms/40 ms break/make ratio working at 10 pulses per second (pps). As mechanical parts wear, the dial's performance will sometimes shift away from these specifications. Because it is critical that properly formed dial pulses are received at the central office, any significant variation in a rotary dialer's operation could cause the CO to ignore or misinterpret the dial pulses being generated.

Unplug your telephone from its wall jack and expose the rotary dial unit. Closely inspect the rotary dial mechanism. Remove any accumulations of dust or dirt using a canister of compressed air. You can also brush away dust or dirt with a very soft-bristled brush. Avoid exerting force on the mechanism—that could bend electrical contacts or shift a mechanical alignment.

Once the mechanism and contacts are clean, add several drops of light, general-purpose household oil into the main rotating shaft of the dial assembly. Work the rotary dial until the oil is distributed evenly into the assembly. Do not use excessive amounts of oil, and avoid heavy oils or grease. Excessive or heavy lubricants will only attract more foreign particles—and cause more problems in the future. (Never apply oils to the electrical contacts.) If the performance of your rotary dial does not improve, the entire assembly should be replaced.

Symptom 5 DTMF keypad is unresponsive. No tones (or only one tone) is generated when a key is pressed, or the key must be depressed firmly to engage both tones. (Classical DTMF dial units only).

A DTMF dial unit uses seven frequencies in combinations to represent twelve unique digits, so each key that is pressed on the dial pad must actuate two oscillators (one oscillator for the row frequency, and one oscillator for the column frequency) simultaneously. When a key is pressed, a mechanical linkage attached to the corresponding key closes the switch contacts that activate the desired oscillators. Dust and dirt can accumulate on these switch contacts and inhibit their operation. Normal metal corrosion can also affect normal switch reliability.

Unplug the telephone and expose the DTMF dial unit. Remove any accumulations of dust or dirt using a canister of compressed air, or a soft-bristled brush. Contact corrosion can be removed by using a quality electrical contact cleaner. Never use harsh solvents to clean electrical contacts. Solvents will clean just fine, but they can also dissolve the plastics used in dial pad construction. Never coat electrical contacts with oils or grease.

Inspect the dial pad's connections to the network. Each wire from the dial pad should be inserted firmly and completely into the network PC board. Loose or fractured wiring should be repaired. Check each connector on the network PC board to ensure that all soldering points are secure. If the DTMF dial unit continues to malfunction, it should be replaced.

Symptom 6 Low or distorted transmission from the telephone. When a carbon microphone is new, its metal diaphragm is rigid and firmly attached to a capsule filled with carbon granules. Over time, however, natural aging causes the diaphragm to stretch. This loosens the diaphragm's tension, as well as its sensitivity. As the diaphragm loosens, less speech energy is transferred into the carbon capsule to be converted into speech current. Resulting speech signals are therefore lower in amplitude. A loose diaphragm also introduces a certain amount of distortion into the transmitted signal. Portions of the speech network might not be picked up by the microphone, resulting in bland, tinny speech.

Distortion in a carbon microphone can also be due to diaphragm corrosion or a buildup of foreign matter. Particles in everyday breath vapor, cigarette and cigar smoke, or airborne dust can accumulate on a diaphragm and inhibit its vibration.

Your best course of action is to simply replace the suspect carbon microphone. A fresh microphone often works wonders for your transmitted signal.

Symptom 7　Low or distorted reception from the telephone. Check your telephone's reception by comparing its reception to that of other similar telephones. If the same signal sounds clear and crisp on other telephones, but weak and distorted on the suspect telephone, the suspect telephone is probably at fault. If all telephones are suffering from a poor receive signal, the fault could be in the central office circuitry, or in the caller's transmitted signal.

If you determine that the suspect telephone is at fault, try replacing the electromechanical receiver element. A receiver uses a diaphragm wrapped with a voice coil of thin wire. Over time, the diaphragm can loosen or stretch. Such aging effects will eventually degrade the fidelity of sound reproduction. The natural accumulation of dust and debris in the receiver area also contributes to sound distortion.

Also, check the network PC board to be sure that connections to the handset cord are secure. Check the PC board for any cracks or loose soldering connections that might interrupt the circuit. If a new receiver does not rectify the problem, replace the network PC board.

7

Electronic telephones

ELECTRONIC TELEPHONES, SUCH AS THE ONE SHOWN IN FIG. 7-1, MAKE UP A much broader category of telephones than the classical variety covered in chapter 6. By taking full advantage of integrated circuit technology, a complete telephone circuit can now be assembled on a single IC chip. Advanced features and functions that would have been impossible with classical telephone technology can usually be offered with little or no additional cost to the telephone buyer.

For the purposes of this chapter, an electronic telephone is any telephone that utilizes at least one IC. In this chapter, you will learn how passive electromechanical devices have gradually been replaced by discrete integrated ICs. IC technology has now been fully integrated in telephone design—making advanced features possible and affordable.

Implementations of ICs

Integrated circuits were first utilized as direct replacements for classical assemblies such as ringers, dialers, and speech networks. ICs generally provide three major advantages to telephone manufacturers. They are:

1. Much smaller and lighter than their electromechanical counterparts,
2. Less expensive and easier to use and manufacture than the electromechanical assemblies they replace.
3. Typically more reliable over the long term than electromechanical assemblies.

It did not take long at all for ICs to prove themselves in the telephone industry.

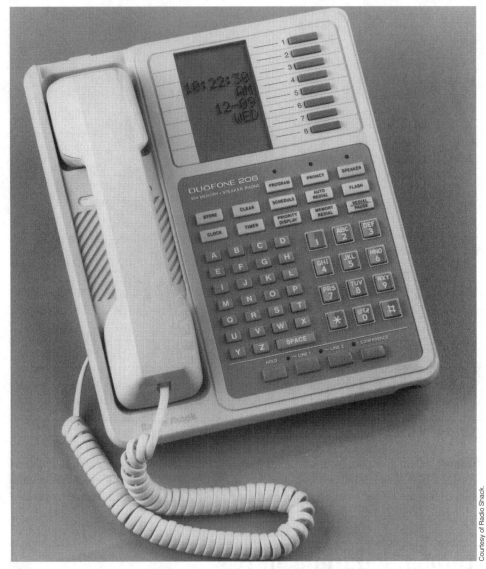

7-1 An advanced electronic telephone.

Ringer ICs

The ringer has historically been the largest and heaviest single part of any classical telephone. Manufacturers welcomed the chance to replace the electromechanical workhorse with a small, light IC that could be added right to the network PC board. With ICs, manufacturers were able to design much smaller and unusually packaged telephones—without the bulk and weight of solenoids and gongs. Although electronic ringers might not be as loud as their mechanical predecessors, many telephone users preferred the more subtle tones over the insistent clang of metal gongs.

Ringer ICs initially had trouble with their power compatibility. Like classical ringers, ringer ICs had to receive power from the telephone line, but ICs operate with low-voltage dc power, while ringing signals supply high-voltage ac power. To produce a fully telephone-line-compatible ringer IC, four major sections were needed: a voltage rectifier, an anti-tinkle circuit, a tone generator (or oscillator) circuit, and an output driver circuit. A block diagram of a contemporary IC ringer is illustrated in Fig. 7-2.

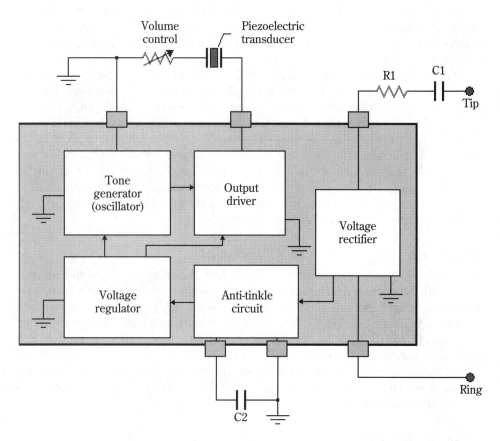

7-2 A typical ringer IC.

Telephones with ICs must be designed to work properly from the wide range of voltages available on a telephone line. A *bridge rectifier* is used on the input portion of the ringer IC to ensure that the ac ringing signals supplying power to the IC will be converted to dc before being applied to the IC's circuitry. A bridge rectifier also ensures that dc voltage polarity within the ringer IC will remain constant regardless of the voltages swinging across tip and ring. External capacitor C1 blocks dc from flowing into the ringer—only ac ringing signals are allowed to pass. The series resistor R1 helps to limit any current flowing into the IC.

Once voltage is rectified, any high-voltage transients on the telephone line, such as the voltage spikes produced by rotary dial contacts, must be removed. If transients are allowed to energize the ringer, they will cause the ringer to chirp or flutter. In an electromechanical ringer, this condition produces a light tinkling sound. An *anti-tinkle* circuit is employed to quench any transients from energizing the ringer. A zener diode and capacitor C2 normally form the anti-tinkle circuit. The zener diode's breakdown voltage sets the level at which quenching will take place, while the capacitor absorbs the transient energy that exceeds the zener's level.

Dc voltage must be regulated before it powers the tone generator or output driver. Regulation prevents fluctuations in the telephone's line voltage from adversely affecting the ringing signal. A regulator circuit often provides more than one regulated output to various locations in the ringer IC.

The tone generator portion of a ringer IC is the circuit that generates the audible sound that will be emitted from the telephone. Tone generators are often categorized as single-tone or multi-tone circuits. Some tone generators are even able to produce music.

The signal created in the tone generator must be amplified by the output circuit, then supplied to the device that creates the audible output. In most applications, a piezoelectric transducer is used to make sound, although some applications use a specially-coupled speaker. An adjustable resistor can be added to the circuit to change the ringer's output volume. The transducer itself is little more than a slab of piezoelectric crystal. When voltage is applied across the crystal by the output circuit, the transducer will vibrate and produce sound.

Dialer ICs

Electronic dialers received almost immediate acceptance by telephone manufacturers. DTMF dialers based on integrated circuits could be produced with a small fraction of the weight and size needed to manufacture dialers using discrete resonant circuits. An IC tone generator is able to produce tones to a much tighter tolerance than is needed by a central office, so cheaper external components (like resistors and capacitors) can be used with the IC.

IC pulse dialers are also available. With an electronic pulse dialer, the mechanical rotary dial can be eliminated completely, and a DTMF-type keypad can be used instead of a rotating wheel.

IC pulse dialing

In order to successfully replace a mechanical rotary assembly, an electronic pulse dialer must be able to accomplish two functions. The IC pulse dialer must first be able to interrupt loop current at the desired rate and break/make ratio. Second, the dialer must be able to mute any receive audio to prevent loud clicks from being heard during dialing. The IC pulse dialer assembly can be interfaced to a telephone as shown in Fig. 7-3. In actual practice, the pulse and receive mute contacts are replaced with conventional bipolar transistors within the IC itself.

The block diagram for a simple pulse dialer IC is illustrated in Fig. 7-4. Five major subsections are required to form a working IC: a key decoder circuit, a memory circuit, a master timing clock (or oscillator), a control circuit, and pulse output driver circuits.

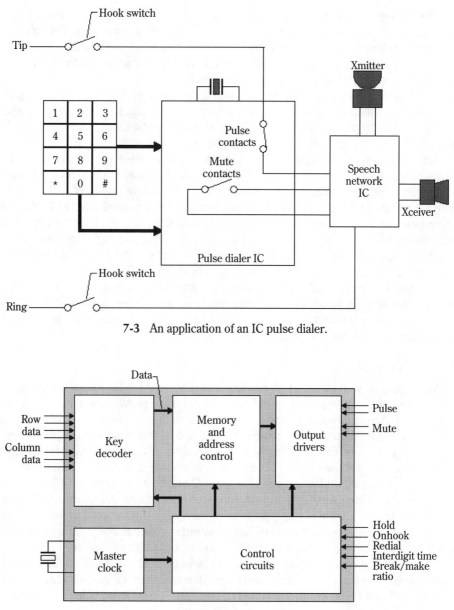

7-3   An application of an IC pulse dialer.

7-4   A pulse dialer IC.

A keypad, such as the one used in DTMF dialers, produces two individual logic signals; a *key row* signal, and a *key column* signal. The combinations of these two signals are unique for each key. A *key decoder* circuit is used to interpret the active row and column signals, then generate a corresponding logic code for that key. This data is stored directly in a temporary memory circuit. Memory is essential, because keys can be pressed much faster than the IC output can pulse.

Two output driver circuits are needed from the IC: a mute circuit and a control circuit. The mute circuit harmlessly short-circuits the receiver during dialing to prevent the receiver from developing loud clicks each time loop current is broken. Of course, a set of pulse outputs is needed to interrupt loop current. Both of these output circuits are usually very simple transistor drivers.

Control circuits are responsible for managing the overall operation of the pulse dialer by coordinating the key decoder, memory, and output driver. Several inputs are included with the control circuits. Typical dialer control inputs are: hold, on/off hook, redial, interdigit time, and break/make ratio.

The *hold* control line can be used to disable pulses from the IC's outputs. The *on/off hook signal* tells the dialer when a telephone is active. The on/off hook signal is sometimes referred to as a "call request" signal. A *redial trigger* is often included to take advantage of the IC's memory. When a redial is requested, the pulse dialer IC will repeat the last number dialed. The *interdigit time* (IDT) setting is the time delay that occurs between each digit (or set of pulses). IDT can range from 0.2 s, to as much as 1 s, but the state of the IDT input usually causes the IC to switch between two fixed delay values that were preset somewhere within that range.

Finally, the break/make ratio can select either of two preset ratios. Normally, the ratio is 60/40 (break/make), but this ratio can sometimes be switched to 67/33. The ratio difference might seem insignificant, but the break/make ratio is one of the most important parameters of pulse dialing.

IC DTMF dialing

DTMF dialing is well-suited to IC applications. Discrete LC (inductor-capacitor) oscillators can easily be replaced with digital oscillators to provide the row and column tones. A DTMF dialer IC must also supply a mute signal to reduce tone volume at the receiver. Figure 7-5 illustrates a simple DTMF dialer interfaced to a telephone line.

There are many variations of IC dialer designs. A block diagram for a typical DTMF IC dialer is shown in Fig. 7-6. Dc voltage to power the IC is normally provided by a rectifier located on the speech network PC board. More intricate DTMF dialer ICs provide a built-in rectifier and regulator—to ensure a small, modular telephone circuit. All DTMF dialer ICs require a minimum of five functional areas to form a working IC: a master oscillator, a key decoder/pulse generator circuit, high- and low-frequency DA filters, a summing amplifier, and an output driver circuit.

In order to produce a stable tone, there must be some sort of time base for the dialer circuit to work from. This *master oscillator* supplies a fixed, extremely stable frequency signal (usually 3.514 MHz). The oscillator's frequency is locked in using a piezoelectric crystal.

A *key decoder/pulse generator* circuit accepts the row and column logic inputs from a keypad. The decoder then divides the master oscillator's signal to produce a high-frequency and a low-frequency digital pulse train. At this point in the process, signals are digital pulses—not audible tones. To produce actual tones, each digital pulse train must be converted to an analog signal.

A *digital-to-analog converter* (DA) changes digital pulses into a corresponding analog voltage that will vary smoothly over time. The converted voltage is then fil-

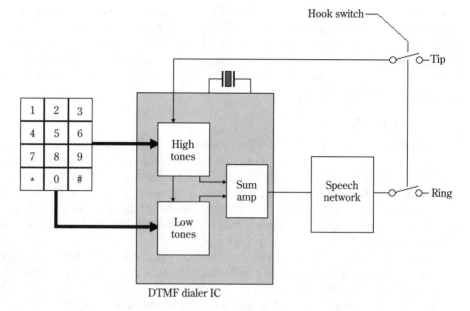

7-5   An application of a typical DTMF dialer IC.

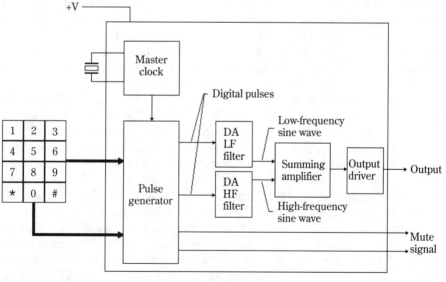

7-6   A DTMF dialer IC.

tered to ensure that only the desired frequencies are passed to the IC's output. A separate DA converter is required for high- and low-frequency pulse trains.

A *summing amplifier* is added to combine high- and low-frequency tones into a single signal that can be introduced to the speech network. An *output amplifier* then boosts the DTMF signal and matches the IC's output to the speech network's input. The speech network then delivers the DTMF signal to the telephone network. A mute output from the dialer is used to inhibit the receiver each time a key is pressed. Otherwise, dialed tones would sound very loud in the receiver. Even though the receiver is muted during dialing, a faint DTMF signal can still be heard—this lets you know that the dialer is working.

Speech network ICs

Speech circuits were the last portion of a telephone to be integrated into an IC chip. Integration had to wait until an efficient, high-gain, low-noise amplifier was developed that could work well from variable telephone line voltages. Development of the operational amplifier finally made speech circuit integration possible, while eliminating the bulk and weight of hybrid transformers. A highly simplified block diagram of a telephone speech circuit IC is shown in Fig. 7-7. Each portion of the IC serves an important function.

In order to replace a discrete speech network, a speech IC must be able to handle a variety of tasks. The IC must be able to channel and amplify transmitted speech, and provide adequate sidetone. A speech IC also must be able to interface a dialer's signals to the local loop, while presenting a constant load on the loop regardless of loop conditions.

A *dc interface* circuit sets the operating parameters for the telephone based on the amount of loop current that is available. An external resistor and capacitor are used to set the IC's operation. A regulator accepts dc voltage from the telephone line, and provides a constant voltage output to power the IC's speech amplifiers. The regulator shown in Fig. 7-7 also supplies a secondary voltage output that can power a compatible DTMF dialer IC. An external capacitor is added to stabilize the regulator's output level.

There are two primary amplifiers in a speech IC: a *transmit amplifier* and a *receive amplifier*. Amplifiers not only increase the signal levels into and out of the speech IC—they can also be used to mix various signals together. A transmitter (either a carbon, electrodynamic, or electret microphone) accepts low-level speech signals that are boosted by the transmit amplifier (At). The amplified signal is then split—a portion of the transmit signal feeds a sidetone amplifier (As), while the majority of the transmitted signal is returned to the regulator. Once it arrives back at the regulator, a speech signal is coupled back into the telephone loop.

The sidetone signal is controlled through a simple filter called a *sidetone balancing network*, then added to the normal received signal feeding the receive amplifier (Ar). An additional amplifier, called an *equalization amplifier* (Ae), is used to continuously adjust transmitter biasing for any variations in loop current. This kind of automatic compensation technique improves a microphone's performance regardless of the loop length.

Although a carbon microphone is often used with an IC speech circuit, solid-state amplifiers permit the use of electret and electrodynamic microphones as well. Due to the inherently high-impedance of an electret microphone and the low impedance of an electrodynamic microphone, it is necessary to add appropriate bias-

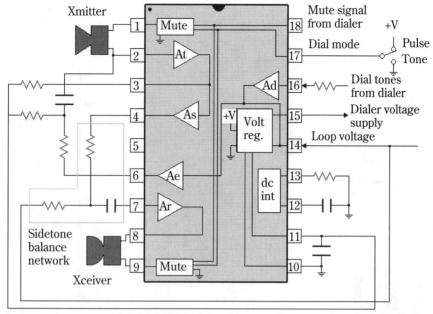

7-7 A typical speech network IC.

ing circuits that properly match the microphone to the IC. A high-impedance microphone interface scheme is illustrated in Fig.7-8, and a low-impedance interface version is shown in Fig. 7-9. Notice that the electrodynamic interface employs a discrete transistor amplifier to pre-amplify the microphone's signal and alter the signal impedance to match the speech IC.

Many speech networks, such as the one shown in Fig. 7-7, support rotary or DTMF dialers directly. A digital logic input is supplied to the speech network to specify the desired operating mode of the telephone. The mode is usually either pulse (rotary) or tone (DTMF). When the DTMF mode is selected, any generated tones are accepted into the speech IC, then amplified by a dialer amplifier(Ad). The amplified tone output is coupled directly to the local loop. A mute signal produced by the dialer is used by the speech IC to shut down both the receiver and transmitter amplifiers. However, the unbalance condition set by the balancing network allows some of the dialing tone to be faintly heard in the receiver.

An electronic telephone

An electronic ringer, dialer, and speech network can be easily combined to form a complete, solid-state telephone similar to the diagram shown in Fig. 7-10. A solid-state telephone can supply features such as multiple ringing tones with an easily adjustable volume control, switchable tone or pulse dialing operation, memories that hold frequently dialed telephone numbers, automatic redial of the last dialed number, and a speech circuit that is free of bulky transformers or coils.

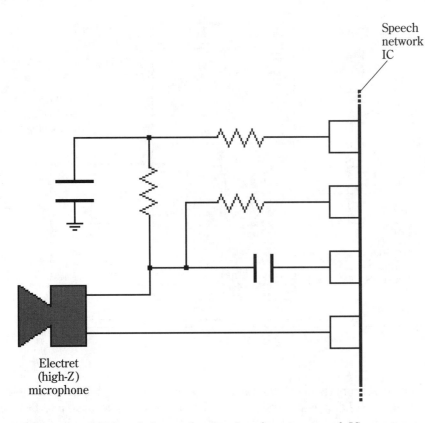

7-8 Interfacing an electret microphone to a speech IC.

The solid-state telephone is powered entirely from the telephone line. A conventional diode bridge rectifier provides ac voltage rectification and voltage polarity protection. A high-voltage zener diode serves to protect the dialer and speech ICs from voltage transients on the telephone line. A ringer IC is connected directly across tip and ring through an RC circuit—the capacitor blocks dc so that the ringer IC will not present a load to the central office unless a ringing signal is present.

The IC dialer can operate in either the tone or pulse mode, depending on the setting of a single selector switch on the telephone. When switched in the tone mode, DTMF signals are coupled directly to the speech network IC. In the pulse mode, a discrete pulsing circuit (operated by the dialer) is used to interrupt the loop current. The interdigit time and break/make ratios are predetermined by the particular dialing IC. A secondary zener diode quenches any high-voltage transients caused by dial pulsing.

The speech network IC processes all transmitted and received speech, as well as DTMF signals. By using discrete impedance matching circuits, the speech IC can be configured to work with carbon, electrodynamic, or electret microphones. An electrodynamic microphone is usually standard. Passive components around the speech IC work to bias the microphone, set the IC's load on the telephone line, and balance

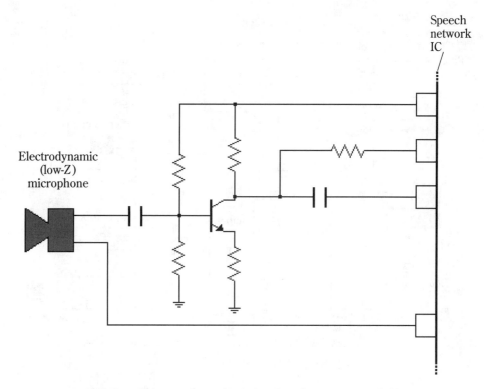

7-9   Interfacing an electrodynamic microphone to a speech IC.

sidetone to the receiver. The entire circuit can be fabricated on a PC board of only a few square inches.

Of course, this is only one of many possible configurations, but it demonstrates the basic concepts and approach needed to implement a fully electronic telephone.

Advanced implementations of ICs

Telephone electronics has kept pace with the ongoing advances in integrated circuit technology. ICs are now available that include an electronic ringer, dialer, and speech network on the same IC—as well as an assortment of other advanced features. This part of the chapter examines such advanced ICs and their supporting electronics.

The integrated telephone

Combining a dialer, ringer, speech network, and dc loop interface onto the same piece of silicon is no easy feat, but as the block diagram of Fig. 7-11 illustrates, it is a remarkably simple and elegant alternative to the individual ICs shown in Fig. 7-10. Each functional part of the IC performs similarly to the individual ICs discussed earlier.

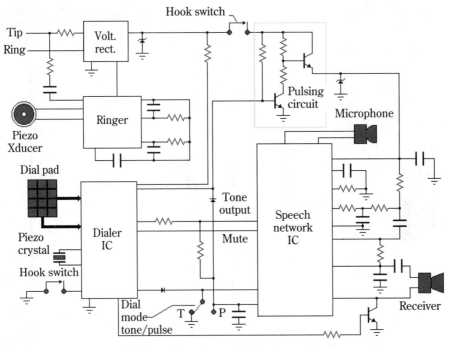

7-10   A complete electronic telephone circuit.

One function, however, is new to some models of integrated telephone IC—a microprocessor interface. Adding a microprocessor to a telephone might seem rather extravagant at first glance—each function necessary to implement a complete telephone can already be fabricated onto a single IC. The telephone itself performs no mathematical or logical computations during its operation.

A microprocessor does nothing to enhance the performance of a telephone's basic functions. The strength of a microprocessor lies in the many added features that it can offer. A microprocessor can be used to support such features as expanded number memory and recall, a visual display (using either LED or LCD technology), a visual calendar, clock, call time elapsed timer, calls waiting, or calling telephone number display (using the Caller ID service now being provided by some BOCs across the United States). A microprocessor can also manage features like outgoing call restriction, call logging, automated redial, and answering system control. Telephones that utilize a microprocessor are often referred to as *intelligent telephones.*

As with individual telephone ICs, integrated telephone ICs require a number of discrete external components to configure their operating parameters such as loop resistance, sidetone balance, and speech amplifier gain. Discrete components also provide voltage polarity and transient protection. The exact number of discrete components needed, as well as their individual values, depends on your particular telephone's design.

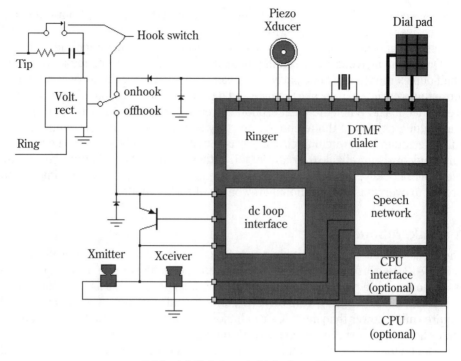

7-11   A fully-integrated telephone IC.

The speakerphone

Speakerphones (also referred to as "comfort" or "hands-free" telephones) are certainly not new devices. They have been in existence in one form or another for several decades. Speakerphones were originally developed for the business community in an effort to relieve the physical discomfort suffered by individuals who spent a great deal of time on the telephone. The use of speakerphones also freed the hands of those individuals for increased productivity.

Speakerphones have a receiver and transmitter built into the telephone's main housing—thus eliminating the need for a handset (although a handset is almost always added for privacy). When a transmitter and receiver are mounted into the telephone, it is possible to speak or listen to a caller from just about anywhere in a room. Speakerphones make it possible for a group of people to listen and participate in the same conversation simultaneously without the need for separate telephone extensions.

Unfortunately, speakerphones suffer from several inherent drawbacks that have until only recently kept them largely out of the commercial market. The speakerphone's greatest disadvantage is the effect of feedback. Output signals from a speaker travel freely through surrounding air. Such speaker signals are usually strong enough to be picked up again by a nearby microphone. The self-oscillation produced by such feedback causes an incredibly irritating wail or whining sound. Feedback does not occur in a handset because there is not enough amplitude from the handset's receiver to stimulate the handset's microphone. Also, the presence of your head against the handset blocks any speech signals between the transmitter and receiver.

Due to the unavoidable possibility of feedback, a speakerphone must operate in a half-duplex mode—only one party can speak at any one time. Circuitry is needed to sense the presence of a speaker's voice and disengage the corresponding receiver to prevent feedback. When the speakerphone is receiving, circuitry must disengage the transmitting portion of the telephone. In early speakerphone models, complex and bulky detection and switching circuitry was required. Not only was such control circuitry hideously unreliable over any length of time—it added tremendous cost to the telephone. Luckily, the new generation of speakerphones use advanced integrated circuits to overcome the traditional drawbacks of earlier models.

A speakerphone IC

The speakerphone IC must contain a number of important circuit areas as illustrated in Fig. 7-12. Because a caller can be located at almost any point in a room, it is necessary to provide a powerful transmit signal amplifier (At), as well as a substantial receiver amplifier (Ar) that can drive a conventional speaker. A speaker usually requires more power than most ICs can handle, but a separate amplifier circuit or IC can be added to boost the speakerphone's output.

A detector circuit must check for the presence of transmitted and received speech, then compare received signals to background noise. In this way, a speech

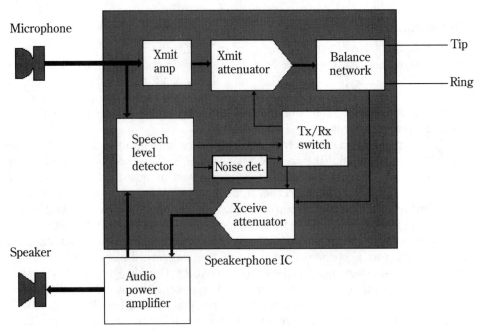

7-12 A speakerphone speech IC.

detector can determine whether the speakerphone should be in its transmit or receive mode. The speech detector and noise detector outputs are interpreted by a transmit/receive switch (Tx/Rx) circuit that controls transmit and receive attenuators. An *attenuator* is a circuit used to reduce (or attenuate) a signal on demand. Both attenuators are arranged in a complementary fashion—one is always off while the other is on.

Finally, the speakerphone IC is interfaced to a speech network connecting the speakerphone to a local loop. A dialer circuit can then be added, along with a selection of discrete components that set the IC's operation and switching parameters. Figure 7-13 shows a block diagram of a complete hands-free speakerphone.

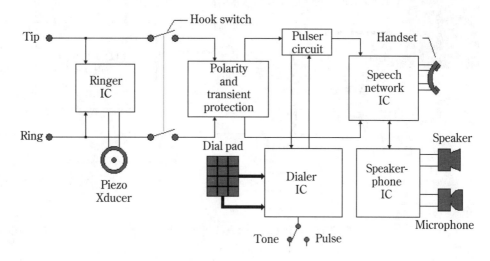

7-13 A complete hands-free speakerphone.

A microprocessor and display

The new generation of electronic telephones offers intelligent features beyond the basic telephone or speakerphone services considered so far in this book. For relatively little additional cost, an alphanumeric liquid crystal display (LCD) can be incorporated into an electronic telephone to provide you with a variety of useful information. Typical examples include called number display, calendar date and current time, call duration, etc.

For an LCD to operate, it must work with a microprocessor. Figure 7-14 shows a block diagram of one typical implementation. Notice that the microprocessor interprets signals from a keypad that includes not only regular dialing digits, but any number of other alphanumeric and control keys. You can see one such implementation of intelligent technology in Fig. 7-11. The microprocessor also directs dialer operation. Instead of sending key information to the dialer directly, keystrokes are first processed through the microprocessor. With microprocessor interaction, it is fairly simple to enter and store telephone numbers for easy speed dialing, along with names, addresses, and other personalized information associated with the telephone number.

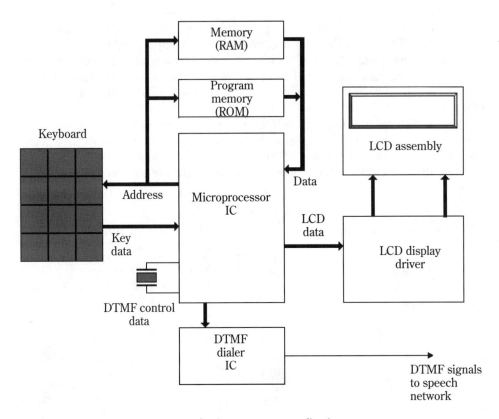

7-14 A microprocessor application.

The microprocessor requires two forms of memory to function. First, some quantity of temporary memory is needed to store the information that you enter (telephone numbers, names, etc.)—and to serve as a scratchpad for any microprocessor operations. Temporary *random-access memory* (RAM) is usually reinforced by using a battery inside the telephone. Second, a microprocessor requires a program—a permanent set of instructions that guides the microprocessor in its operations so it knows how to respond to keystrokes, onhook/offhook conditions, low power, etc. These instructions do not change—they are stored in the telephone's permanent read-only memory (ROM). Depending on just how customized the microprocessor's central processing unit (CPU) is, memory areas can be incorporated right onto the CPU IC itself.

After the telephone's conditions and keystrokes have been interpreted and processed, the CPU sends the appropriate data to an LCD driver circuit. A driver accepts characters and control codes from the CPU, then generates the display timing and control signals that operate an LCD. Depending on the level of sophistication and customization of the particular CPU, LCD driver circuitry might be integrated right onto the CPU IC itself.

Of course, it is certainly possible to combine an integrated telephone IC, speakerphone IC, and CPU/display to produce an intelligent electronic telephone/speakerphone system. Such telephones are among the most sophisticated and expensive devices available to telephone users, but intelligent telephones/speakerphones supply the greatest power and flexibility of just about any personal communication tool available.

All of the additional features discussed in this part of the chapter do not come without a price of power. The circuitry needed to support a microprocessor, memory, visual display, etc., cannot operate reliably with the low currents and voltages typically found on most telephone local loops. Advanced electronic telephones are usually supplemented with an external power supply that converts 120 Vac into a low dc voltage that can be regulated and filtered further inside the telephone. External power sources can greatly reduce the power demands placed on local loop and central office equipment.

Troubleshooting electronic telephones

While the troubleshooting process used with electronic telephones is not terribly complicated, it does require much greater attention to detail than for the conventional telephones discussed in the last chapter. You will undoubtedly encounter a broad variety of circuitry and assemblies that differ with each model and manufacturer. The key elements found in many of today's electronic telephones are illustrated in Fig. 7-15. The figure represents a block diagram for a single-line, electronic telephone/speakerphone with a visual display.

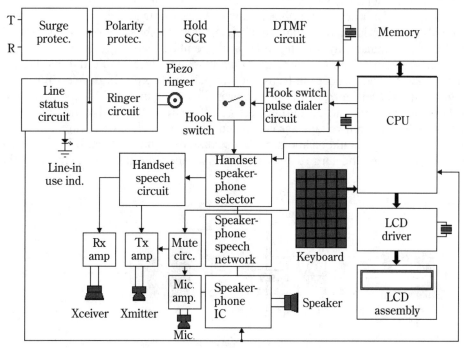

7-15 Troubleshooting block diagram.

Symptom 1　The telephone is completely dead. No sidetone or central office dial tone is audible in the receiver (or speakerphone output). Before you begin to disassemble your telephone, inspect the external power supply (if any) that is supplying it. While many electronic telephones are powered directly from the local loop, almost all electronic telephones incorporating advanced ICs require some form of power supply to support the circuitry—the local loop simply cannot provide enough power at the needed voltage and current levels.

Power supplies typically take the form of small ac/dc converters that convert 120 Vac wall voltage into a low dc voltage that is then plugged into the telephone. Disconnect the power supply from the telephone and use your multimeter to measure the dc voltage at the converter's output. Your reading should be approximately equal to the rated output marked on the converter. Although you will be measuring low-voltage dc, take care to avoid touching the metal conductors at the power connector. Another thing to keep in mind is that in rare instances, converters can be an ac/ac type—providing low-voltage ac to the telephone (a power conditioning circuit in the telephone would convert the low-voltage ac to dc). If your particular converter is an ac/ac type, set your multimeter to measure ac voltage.

Make sure the converter is seated firmly in the ac wall outlet, then re-check your previous measurements carefully. If the converter's rated output is low or absent, the converter might be defective. Try a new power converter.

If your power converter appears to be working properly (or if your telephone does not require an external power supply), check the telephone's line cord next. A line-powered telephone will malfunction if it becomes cut off from the local loop. Try a new line cord or a new line cord in a known good working jack. Line cords are likely candidates to fail due to the pulling, flexing, and general abuse they must endure. Replace or re-terminate any defective line cord assembly.

Should your telephone continue to malfunction, disassemble the telephone's main housing and inspect the transient and polarity protection devices in the circuit. A typical transient/polarity protector is illustrated in Fig. 7-16. Use your multimeter to measure the dc output voltage (V_{out}) from the bridge rectifier. Also measure the dc input voltage across tip and ring at the telephone's input. The output voltage (V_{out}) should be about 0.5 V less than the input voltage. If V_{out} is unusually low or absent, the surge suppressing transformer might be open-circuited, the

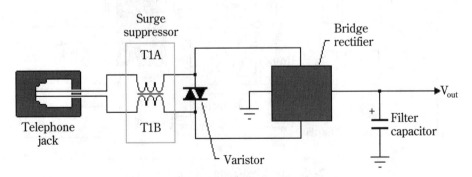

7-16　A transient/polarity protection circuit.

zener diode might be shorted, or one or more bridge diodes might be open-circuited. Disconnect all power from the telephone and inspect each of these components as discussed in chapter 4.

Finally, inspect each of the connectors that tie the telephone together. Make sure that all inter-board connectors are installed correctly. Reseat or tighten any loose connections.

Symptom 2   Speakerphone functions do not work in transmit and/or receive mode. Begin by finding the extent of the malfunction. Check the operation of the handset. If the telephone seems to work properly from its handset, but not from its speakerphone system, the trouble probably lies in the speaker and/or microphone, or some part of their control and amplifying circuitry. If the telephone does not work at all, refer to symptom 1.

Check the speaker and microphone connectors inside the telephone. Gently wiggle the wiring to stimulate any intermittent connections. Repair any faulty wiring and reseat any cables that appear loose.

If there are no obvious connector problems, your best course is to replace the speakerphone IC. A speakerphone IC contains most of the amplifying and switching circuitry needed to operate the telephone in its speakerphone mode. A defective speakerphone IC will easily disable a speaker, or microphone—probably both. If space permits, install an IC socket before inserting a new speakerphone IC. Use of an IC socket prevents you from having to desolder the same points in the future and risk damage to the PC board.

Disconnect all power from your telephone and use your multimeter to test any discrete transistors in the speakerphone switching or control circuits. (Transistor checking is discussed in chapter 4.) An open or shorted transistor can prevent proper speakerphone operation, so replace any defective transistors. Keep in mind that you will probably need a schematic of your telephone to determine which components are associated with the speakerphone circuit.

Some speakerphones use small voice transformers to match and couple signals to and from the speakerphone network. On rare occasions, one of these transformers might develop a short or open circuit that can disrupt a speakerphone's operation. If the telephone has been dropped or otherwise abused prior to its malfunction, one or more of the transformer's leads might have snapped or pulled away from its corresponding PC trace.

Disconnect all power from the telephone and inspect each transformer carefully. Use your multimeter to check each transformer winding. Transformers that appear open or short-circuited should be replaced. If you doubt the meaning of your readings, compare your readings against those taken at other similar transformers. It might be necessary to remove at least one leg of a transformer's winding from the PC board before taking a measurement to prevent other circuit components from causing false readings. Replace any transformer that appears open or shorted.

Symptom 3   The handset does not work in transmit and/or receive mode. Tracking down the source of a handset problem can usually be accomplished fairly quickly. Begin with a new handset cable. Most handset cables used in electronic telephones are modular, so it is a simple matter to exchange the suspect handset cable with one from a known good working telephone. You can obtain new modular hand-set cords from just about any department store with a home electronics department. Handset cable problems are common because of the tremendous amount of bending, flexing, and stretching that such cords must endure. Open or intermittent connections at the modular connectors are typical.

Although most handset cords are directly compatible (after all, there are only four wires), some manufacturers deliberately make their cables nonstandard by reversing one or more pairs of wires. Inspect both the old and new handset cords carefully before you install the new one. Make sure that the wiring in each handset cable is identical.

Another likely trouble spot is the handset assembly itself—especially if the trouble started after the headset was struck or dropped. If the transmitter is not working, suspect the microphone element. If the receiver is not working, suspect trouble in the receiving element. If the handset is totally dead, suspect a wiring problem within the handset.

The body of most contemporary handsets is constructed with two molded halves as shown in Fig. 7-17. A small, hollow upper shell is molded with large, sturdy clips at both ends. These clips interlock with closely molded holes in the lower half of the handset body. After the transmitter and receiver are wired at the factory, both halves of the handset body are snapped together and fastened with one or two screws.

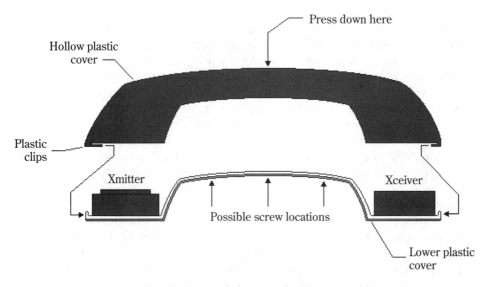

7-17   Opening a typical electronic handset assembly.

Disassembling a contemporary handset is not as simple as it appears at first glance. Remove any fastening screws from the handset (they might be covered by a decorative strip of plastic above the dial pad), then place it, mouth and earpiece down, on a firm surface. A tabletop covered with a soft towel is ideal. With the palm of one hand, exert a firm, constant, downward force at the center of the handset body. This action pushes down on the hollow shell, and pops at least one set of clips from the lower body. Be firm but gentle—too much force can crack the handset! After one set of clips is free, the clips at the opposing end will disengage easily. Do not be surprised if you must make several attempts. Handsets are almost fiendish in their design, so it takes a large amount of force (and a little practice) to separate the handset bodies.

Once the innards of a handset have been exposed, examine the handset's wiring connections. Tighten any loose connections or repair any faulty wiring. Inspect the transmitter and receiver to be sure that they are seated firmly into their electrical contacts. If the transmitter or receiver is not making proper electrical contact, they can behave intermittently—or not work at all. Finally, try replacing the microphone or receiver as appropriate.

Reassembling a handset is often easier than taking a handset apart. Insert the plastic clips at one end of the upper housing into their corresponding holes in the lower housing, then simply press both halves together until all clips lock into place. Replace any fastening screws.

If none of the procedures up to this point restore the operation of your handset, test your hook switch next. Remember that your hook switch effectively connects a telephone to the local loop. If any hook switch contacts remain open, the telephone will remain idle and no part of the handset will function. Check each hook switch contact to make sure that it is not obstructed and is moving freely. Remove all power from the telephone and use your multimeter to check continuity across each contact set. Replace the hook switch if any of its contacts appear to have been bent or broken.

Beyond this point, the trouble is probably in the telephone's speech circuit. Replacing the speech network IC will often correct the problem. You can also inspect any nearby transistors using your multimeter as described in chapter 4. An open or shorted transistor (such as the receiver's mute circuit) can easily inhibit the speech network. Keep in mind that electronic telephones containing both a speakerphone and telephone must have a circuit that can switch back and forth between both modes. If the speakerphone/telephone switching circuit is defective, the speakerphone mode could function, but the handset mode might not engage properly. Depending on the sophistication of your particular telephone, the telephone/speakerphone switch circuit might be integrated onto the speech network IC, or it could be implemented with discrete transistors on the telephone's main PC board.

Symptom 4 The visual display is absent or erratic. There are generally three components that directly affect a visual display's operation: the telephone's microprocessor (or CPU), the CPU's time base crystal, and the display assembly itself. A defect in any of these areas can seriously impair the display.

Begin your check by exposing the telephone board that contains the CPU. For the purpose of this text, leave power applied, but take extra care to prevent any short circuits with either power or signal wiring—the CPU must be running. First, you must find the crystal located next to the CPU (the CPU is typically the largest IC on the telephone's main PC board). A CPU uses at least one piezoelectric crystal to lock in a single, precise oscillator frequency (called a *clock*). The oscillator signal operates the CPU by coordinating the processing of programmed instructions. If the clock fails, a microprocessor will not be able to operate at all. Without a microprocessor to provide regularly updated information, the display driver will simply halt. The display itself might freeze in its current state, it might display meaningless, erratic information, or the display output might just disappear entirely.

You can check a CPU's clock using a logic probe as illustrated in Fig. 7-18. Attach the probe's ground wire to a telephone ground (often at the external dc voltage input), connect the probe's power wire to telephone power (usually +5 Vdc or +12 Vdc available from the external power supply), then touch the metal probe tip to each crystal lead.

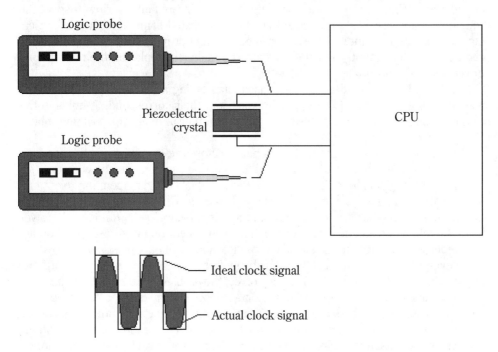

7-18   Checking a CPU's clock signal.

If the clock is working, you will see a pulse or clock indicator light on the probe. If the probe indicates a logic 1 or logic 0, the clock might not be running. To ensure an accurate reading, make sure your logic probe is rated to work with the frequency marked on the crystal. The clock frequency is usually under 5 MHz, and should present no problem at all for most quality, commercial logic probes. If you want an exact visual indication, you can repeat this test using an oscilloscope. An oscilloscope should show a coarse square wave, probably rounded at its corners. If the oscilloscope displays a constant dc voltage, the clock is faulty.

If the clock signal appears to be missing, remove all power from the telephone and try replacing the crystal. Replace the crystal only with another crystal of exactly the same rating—any other value can easily cause erratic circuit behavior (if the circuit works at all).

You might find a second crystal at the LCD driver IC, and perhaps a third crystal at the DTMF tone generator IC. Each of these crystals is used to lock in a clock signal for its particular IC. If your LCD driver is integrated into your CPU, there might be a second crystal at the CPU. Regardless of the number of crystals, the clock signal can be checked with a logic probe or oscilloscope as described above.

A crystal is needed only to fix a desired rate of oscillation. The oscillator circuit itself is integrated into the associated CPU. If a new crystal does not restore a missing clock signal, replace the CPU. If you find the clock signal missing at the LCD driver, replace the driver IC. When the LCD driver is integrated with a CPU on a single IC, the entire IC must be replaced. If space permits, install an IC socket before inserting a new IC. Whenever you are working with ICs, be sure to observe all static precautions to prevent accidental damage to new ICs.

Before attempting to replace any ICs, inspect each connection at the LCD closely. Make sure that any wiring or connectors are tight and inserted correctly. Check for any shorted or broken wires that might have disabled the display. If the telephone has recently been dropped or abused, the main PC board (or a supplementary PC board) might have developed a fracture. Inspect all PC boards and repair if necessary using the PC board repair techniques presented in chapter 3.

After all other possibilities have been exhausted, try replacing the LCD assembly itself. Age or abuse might have caused internal (unseen) damage to the LCD. Be extremely careful when soldering to the new display—excessive heat can destroy LCDs. Also take precautions against static damage to the new LCD.

Symptom 5 Pulse dialing does not work on one or more keys. A rotary dialing error can manifest itself in several different ways. One common error occurs when you press a key and a tone is generated instead—this happens on telephones that have both tone and pulse dialing. Check the pulse/tone selector switch and make sure that the switch is in the correct (pulse) position. If the switch is jammed or very loose, check and replace the damaged selector switch.

Another common problem can arise when one or several dialing keys malfunction, or must be pressed firmly or repeatedly to generate a response. Such symptoms suggest a fault in the keypad assembly itself. Disconnect all power from the telephone, expose the keypad assembly, and carefully inspect any connectors or interconnecting wiring leading to the assembly. Tighten any loose connectors and gently reattach any broken wires. As you can see in Fig. 7-19, an interruption in one of the row or column wires can disable an entire bank of keys. Also examine the keypad PC board and check for any cracks or trace breaks.

The keys themselves might also be defective. Membrane keys are terribly unreliable over long-term use. The flexible membrane used as the electrical contact eventually stretches and breaks, leaving the switch as an open or short circuit. Few membrane switches are sealed, so moisture from the environment can cause contacts to corrode and become unresponsive—an old membrane switch might have to be pressed hard several times before a good contact is made. Chapter 2 provides more information on membrane keys.

While membrane keys are still used in some newer telephones, most manufacturers use contact switches where the electrical contacts are etched right onto a PC board. Figure 7-20 shows a cross-sectional diagram of a typical type of contact switch assembly. The conductive points of a switch are etched onto a thick PC board—a thick PC board will add strength to the assembly whenever a key is

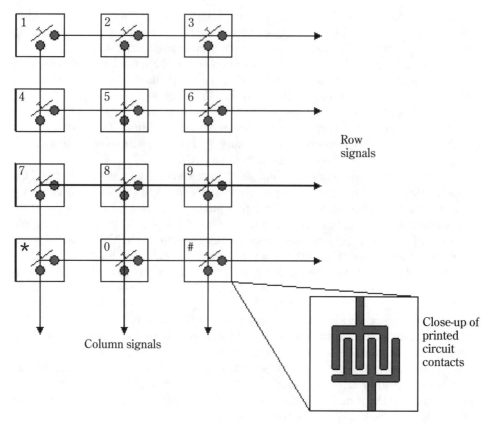

Row signals

Column signals

Close-up of printed circuit contacts

7-19 The configuration of a simplified keypad.

pressed. A conductive element is fitted into a flexible rubber overlay that is placed over the PC board contacts. Each set of contacts has a conductive element positioned over it.

Finally, the assembly is mounted into a molded plastic cover. Each rubber key fits through a corresponding hole in the plastic cover. When a rubber key is pressed, the rubber compresses, and the conductive element presses against the corresponding contacts etched onto the PC board. This action completes the switch. When the key is released, the rubber expands, lifting the conductive element and breaking the switch contact.

Conductive inserts used in newer, better-quality telephone keypads use conductive elements made from a conductive elastomer. Elastomers are highly reliable and enjoy a long working life. Older, less-expensive telephones often use conductive elements made from simple plastic impregnated with carbon fibers. When a key makes contact, carbon is deposited onto the etched contacts. After many cycles, carbon deposits build up on the etched contacts and cause high-resistance connections, and keys have to be pressed harder to make a good contact. Combined with the natural buildup of dust and moisture from the environment, a key can become totally inoperative.

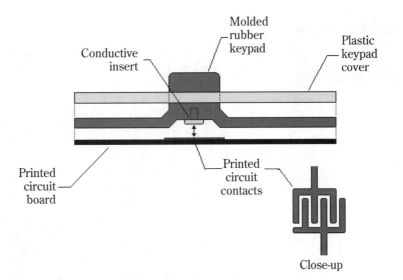

7-20   A printed contact switch assembly.

The solution to faulty contact switch problems is to clean each PC contact thoroughly with a clean cotton swab dipped in ethyl alcohol or quality electrical contact cleaner. The carbon residue should immediately come off. Rub firmly to ensure that all foreign materials are removed.

You should also clean the contact surface of each conductive element to remove any dirt or debris that might have accumulated. Never sand or scrape the contacts of conductive elements with sandpaper or steel wool. Rough cleaning will destroy the PC contacts and conductive elements and only accelerate future problems.

Be extremely cautious of harsh solvents or cleaners that can destroy plastic housings or rubber parts. If you are unsure what effect your particular solvent will have on the plastic housing, test a small area of a housing's inside surface. If the test patch mars or melts, do not use the solvent.

If your pulse dialer is still defective at this point, it is probably safe to presume that either the CPU (or dialing IC) has failed. Use your logic probe to measure the logic output of the dialing IC or CPU as illustrated in Fig. 7-21. Try dialing now. If you read a pulse indication on your logic probe, then your problem is probably in the discrete pulsing circuit. Keep in mind that at 10 pps, some logic probes will show a pulse train as a series of high and low transitions instead of a pulse condition. As long as the CPU or dialing IC is generating pulse control signals, remove all power from the telephone, then use your multimeter to check each of the discrete transistors in the pulsing circuit (the switching network that actually interrupts the loop current). Replace any transistors that appear defective. If no pulse signals are being generated by the CPU or dialing IC, the respective IC is probably defective and should be replaced. If space permits, install an IC socket before you insert a new IC.

Keep in mind that some line-powered telephones operate at voltages too low for a logic probe to measure properly. If your probe readings are inconclusive, use an oscilloscope to measure the pulse logic signals.

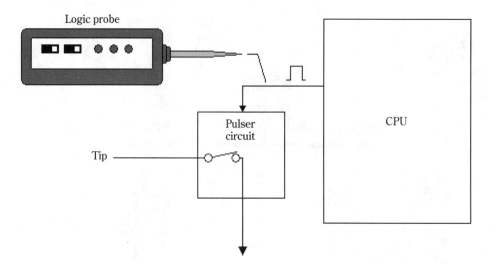

7-21   Measuring a CPU's pulse output.

Symptom 6   Tone (DTMF) dialing does not work on one or more keys.
When an electronic telephone offers a choice between tone or pulse dialing modes, make sure that the pulse/tone selector switch is set for tone mode. If you find the selector switch to be jammed or very loose, check and replace the damaged selector switch.

Disconnect all power from the telephone, expose the keypad assembly, and carefully inspect any connectors or interconnecting wiring leading to the assembly. Tighten any loose connectors, and repair any broken wiring. The keypad assembly itself could be malfunctioning. The keypad PC board might be cracked or broken, membrane switches might be defective, or contact switches might be dirty. Read through symptom 5 for a discussion of keypad maintenance. If the DTMF keypad is entirely defective, a more serious problem could be indicated. Power should be applied to the telephone for the following test.

CPU and DTMF ICs use crystals to precisely lock-in clock frequencies that each device requires for proper operation. If your telephone uses a CPU, use your logic probe or oscilloscope to measure the clock signal at the CPU crystal, as shown in Fig. 7-18. A normal CPU clock signal should appear as a clock or pulse display on the logic probe. An oscilloscope display should reveal a slightly rounded square wave. The CPU needs a clock to coordinate its processing of program instructions. If the CPU clock is missing, replace the corresponding crystal and repeat the check. Also repeat this test for the DTMF IC.

Use your logic probe or oscilloscope to check for the DTMF clock signal. If the DTMF clock is absent, the IC will not be able to produce the tone pairs necessary for dialing. If the DTMF clock is absent, replace the DTMF crystal.

If a new crystal fails to restore the CPU clock, or the DTMF output still does not function, replace the defective CPU (or dialer IC) outright. Use extreme caution when handling and installing new ICs to prevent static discharge from your skin or clothing.

If space permits, install an IC holder before inserting a new IC. Take all static precautions to prevent damage to new ICs (and to the telephone you are repairing).

Some telephones use discrete transistor networks to channel DTMF signals through to the local loop. It is possible that one or more of the transistors might have failed. Remove all power from the telephone, and use your multimeter to check each suspect transistor. Replace any transistors that appear to be defective.

Symptom 7   Ringing is faint or absent. Before you open the telephone, check the ringer volume control (if available). Many electronic telephones include a ringer volume control and/or a switch to turn the ringer on and off. A well-meaning adult or curious child might have reduced or turned off the ringer sound.

Disconnect all power from the telephone, expose the main PC board, then use your multimeter to check the condition of the dc blocking capacitor. (Chapter 4 discusses how to check capacitors.) Replace the capacitor if it appears open- or short-circuited. An open-circuited capacitor will prevent ac ringing signals from reaching the ringer IC.

Use your multimeter or oscilloscope to measure the ac ringing signal reaching the ringer IC's input pin. When a ring occurs, you should read an ac voltage around 90 Vac. If there is no ringing signal available, trace the circuit back to tip and ring. Multi-line telephones might use a ring coupling transformer to couple multiple ringing signals to a single ringing circuit. Check any such transformer to be sure that its windings are intact.

If you can confirm that a valid ringing signal is reaching the ringer IC, check the ringer IC's output. A ringer's output is usually connected directly to a piezoelectric buzzer. If the ringer IC's output is absent, replace the ringer IC. If you are using an integrated telephone (ringer, dialer, and speech network combined onto a single IC), the entire IC must be replaced. If there is a valid output from the IC, replace the piezoelectric buzzer.

Symptom 8   No redial or number storage functions. Often, trouble with the redial or number memory can be traced to a problem in the keypad. If your keypad has been acting erratically, or if particular keys seem to be defective, check your keypad against the procedure in symptom 5. Also check for faulty wiring or loose connectors that might interrupt the keypad signals.

If your telephone uses a CPU, the CPU itself could have failed. A CPU failure can manifest itself in areas such as dialing and display problems. A defective CPU might not be able to access its memory areas where telephone numbers are stored. Try replacing the CPU. Use extreme caution when handling a new CPU to prevent static discharge from accidentally damaging the device. If space permits, install an IC holder before you insert the new CPU.

If your telephone does not have a CPU, or the CPU is not at fault, the dialer IC has probably failed. Try replacing the dialer IC. Take all static precautions to prevent accidental damage to the new IC, and install an IC holder (if possible) before inserting a new IC.

Finally, try replacing the number storage memory (RAM). Most newer telephones integrate the number memory into the CPU or dialer IC.

8

Answering machines

IT IS HARD TO IMAGINE THAT ANSWERING MACHINES WERE ONCE CONSIDERED to be luxury items affordable only by the largest businesses. Today, answering machines are a virtual necessity for all types of businesses and are commonplace in most homes. Answering machines make widespread use of ASICs (application-specific integrated circuits) and microprocessors to handle call processing. High-level integration is the primary reason for the substantial reduction in answering machine size and cost over the last decade (Fig. 8-1). The use of complex, customized ICs has also enabled manufacturers to offer a suite of sophisticated features that simply were not possible on earlier machines.

Now that answering machines have gone from being stand-alone devices to being fully integrated with many new telephones, no text on telephone troubleshooting and repair would be complete without at least one chapter devoted to those machines you still hate to talk to—answering machines.

Conventional answering machines

The first type of answering machine to be covered is the stand-alone (or conventional) answering machine. A conventional answering machine uses one or two magnetic tapes to hold incoming and outgoing messages, as shown in the block diagram of Fig. 8-2. Although the figure might appear daunting at first glance, don't let it fool you. Many of the functional areas shown can be implemented with only a few ICs and some basic components.

8-1   An advanced answering machine system.

Machine functions

Figure 8-2 is probably easier to follow if you consider some of the various functions that a typical answering machine must be capable of performing. First, an answering machine must be able to detect the presence of a ringing signal. While some machines offer an audible ringer for the sake of convenience, an audible ringer is not necessary for proper operation.

A common ringer circuit—very similar to the ringer circuits used in electronic telephones—supplies the incoming ring to a ring detector circuit. In turn, the ring detector circuit converts analog ringing signals into logic conditions that a CPU (or ASIC) can count. The CPU compares the number of rings to the desired number of rings set by a *ring selector switch* on the answering machine. Many machines offer two rings or four rings as common settings.

Once the CPU determines that an appropriate number of rings has occurred, the CPU orders a relay to engage and connect a speech network in the machine to the local loop. This line seize relay acts as a microprocessor-controlled hook switch. Once a speech network is engaged, the answering machine draws loop current. Your local central office detects that your line has picked up, and the incoming call is connected.

As with a telephone, a speech network—almost always in an IC form—is used to interface a two-wire telephone loop to a four-wire circuit (two wires for transmitting, two wires for receiving). The caller's voice is amplified and made available at the answering machine's speaker.

If you are home, a remote speaker allows you to listen to the caller without picking up the call—a feature known as *call screening*. The speech network is also connected to the play/record (P/R) amplifier so the tape system can play to or record from the telephone loop. An amplified microphone is added to allow the recording of new outgoing messages as desired.

Note the presence of a DTMF decoder IC. Older answering machine models used hand-held remote controls to trigger a machine to reset or play back messages.

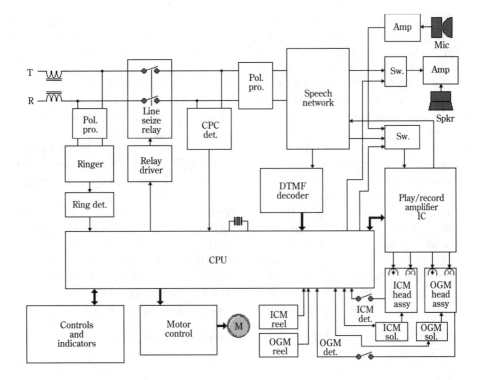

8-2 A stand-alone answering machine.

A DTMF decoder allows nearly all of a machine's available functions to be controlled from any DTMF telephone. Where a DTMF dialer IC uses logic signals from a keypad to generate a set of precise tones, a DTMF decoder works in just the opposite fashion—the high and low frequencies of any DTMF signal are precisely filtered to generate a unique logic output. A CPU can interpret the pattern of logic signals, and determine what actions should be performed. This feature is called *beeperless remote control.*

Another unique answering machine circuit, the *calling party control* (CPC) circuit, takes advantage of the *CPC pulse.* A CPC pulse is generated on the called local loop if the caller hangs up. The CPC pulse creates a momentary interruption in current on your loop that is electrically similar to a dial pulse but has a different duration. The answering machine's CPC circuit detects this interruption and allows the machine to disconnect as soon as possible after the caller hangs up.

There are two standard CPC pulse durations: short (about 10 ms), and long (about 350 ms). A *CPC selector switch* is often provided to allow the answering machine to operate with long, short, or no CPC pulses. The CPC pulse duration varies from CO to CO—some central office facilities do not supply CPC pulses at all.

After the CPU seizes the telephone line, the outgoing message (OGM) must be played. The CPU (or ASIC) sends a control signal to a motor driver and starts the OGM tape. The CPU then sends a control signal that fires a solenoid to engage the OGM play/record and erase heads. A position switch returns a logic signal to the CPU that indicates whether the OGM solenoid (carrying the OGM play/record and erase heads) has engaged properly. An OGM reel sensor tells the CPU that the OGM tape is in motion. If either sensor suggests a problem, the answering machine's cycle aborts.

How a particular machine handles trouble depends on the program instructions stored in permanent memory. If a new OGM is being recorded instead of played, speech from the on-board microphone is amplified, channeled through the speech network, switched through the P/R amplifier, and sent along to the OGM play/record head. The OGM erase head (physically located just ahead of the play/record head) is activated to erase any previous OGM during recording. When the OGM is played back, the message is picked up by the OGM play/record head, boosted by the P/R amplifier, then interfaced to the local loop through the speech network.

When the OGM is finished, the OGM mechanism is released and the incoming message (ICM) solenoid (carrying the ICM play/record and erase heads) is engaged. Motor force now drives the ICM tape. A position switch tells the CPU when the ICM mechanism has engaged properly, and an ICM reel sensor indicates whether or not the ICM tape is moving. If either sensor indicates a problem, the machine's cycle will abort. An erase head is activated during recording to clear any previous ICMs.

An ICM cycle usually continues until one of four conditions occurs:

1. The ICM tape is exhausted and the ICM reel sensor indicates that the ICM tape has stopped.
2. The preset recording time (i.e., 1 minute, 2 minutes, 4 minutes, etc.) has elapsed.
3. The caller stops talking for more than a few seconds in voice-operated control (VOX) mode.
4. The caller hangs up and the CPC pulse is detected.

The motor itself cannot operate directly from a CPU. A motor-driver IC, usually accompanied by an array of discrete transistors, is employed to convert the CPU's logic signals to the voltage and current levels needed to operate a motor.

Finally, all answering machines offer a variety of controls and indicators. Typical controls include the obvious tape controls: playback, rewind, fast-forward, stop, erase/reset, etc. However, there are often more subtle controls: CPC selection, ring count selection, OGM recording controls, and machine-mode controls such as answer only, toll saver, or memo. An answering machine has several displays. As a minimum, it displays such things as machine-in-use, power, and the number of calls that have been recorded. Your particular machine might have other controls and displays. Now that you have seen an overview of an answering machine, you can take a closer look at some of its circuitry.

Answering machine circuitry

Like the telephone, an answering machine uses a mix of electronics and mechanics to accomplish its functions. In this part of the chapter, you will learn about the machine's electronics in more detail. Many of the design principles used with answering machines are very similar to those used with telephones.

Ring detection circuit

An answering machine must be able to detect the presence of a ring signal from the local central office, and pass that ring signal safely to the answering machine's CPU. Ring detection is often handled with a circuit such as the one shown in Fig. 8-3. The ring signal occurs across tip and ring, and is passed by a dc blocking capacitor. A pair of zener diodes provide bipolar (positive or negative) voltage transient protection. An optoisolator is turned on during half of each ring cycle which, in turn, fires an associated transistor. The LED transmitter and phototransistor receiver are completely isolated from each other, so there is no electrical connection between the high-voltage ac ring signal and the sensitive, low-voltage CPU. The output signal from the phototransistor is pulled up to a safe logic level (+5 Vdc) using a resistor.

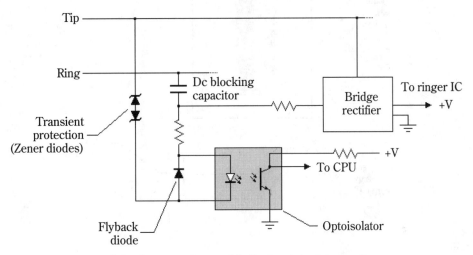

8-3   An answering machine's ring detection circuit.

As the ringing signal alternates polarity, the optoisolator turns on and off. These transitions produce a *digital pulse train* at the CPU. The CPU counts each pulse, then determines when the line should be seized. If the answering machine uses an internal ringer, there will be additional polarity protection circuitry supplying the rectified ringing signal to the ringer IC.

Line seize

Because an answering machine must seize a telephone line automatically, a relay is used, as illustrated in Fig. 8-4. The CPU produces a digital signal that operates a relay driver switch. In most cases, the driver is little more than a transistor configured as a switch. A logic level from the CPU activates the switch and fires the relay. Once relay contacts close, the answering machine is connected to the local loop and draws loop current. The central office interprets the line's offhook condition and connects the calling party. After a recording cycle is finished, the CPU removes the seize signal. This

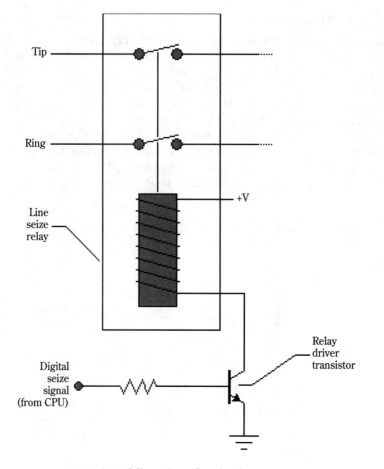

8-4 A line seize relay circuit.

condition turns off the relay driver switch and allows the relay contacts to open, thereby disconnecting the answering machine from the local loop.

Calling party control

An answering machine should disconnect from the local loop as soon as a caller hangs up. Immediate disconnection saves quite a bit of recording time that would otherwise be wasted. Many central office facilities offer calling party control (or CPC) signalling. When a caller hangs up, a brief interruption in loop current is produced on the called line. A circuit such as the one shown in Fig. 8-5 detects the CPC pulse and supplies a corresponding logic signal to the CPU.

As long as current flows in the local loop, the CPC optoisolator is kept in the on condition. The activated LED transmitter, in turn, activates the transistor that produces a logic output for the CPU. When a caller hangs up, the CPC pulse interrupts loop current, allowing the optoisolator to turn off briefly. As a result, a short logic pulse is generated at the CPU. If the interruption is for the proper duration (usually

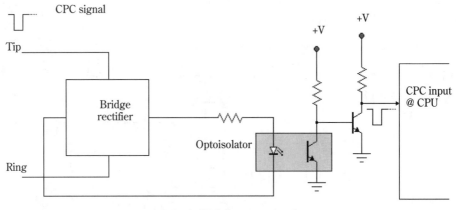

8-5 A CPC detector circuit.

10 ms or 350 ms), the CPU determines that the caller has hung up and the answering machine will immediately release the telephone line. The answering machine then resets in preparation for another call.

Remote control

A DTMF decoder is an integral part of any contemporary answering machine's remote control system. Remote control allows the dial pad of any calling telephone (capable of DTMF operation) to become a control panel that can access just about any of an answering machine's functions. Of course, not all remote control circuits offer that much control. Control capabilities vary with the design of your particular machine.

A simplified remote control circuit is shown in Fig. 8-6. The signal from a seized telephone line is applied to an amplifier, then distributed to a set of tone decoder ICs. One decoder is set to detect the low-frequency portion of a DTMF tone, while the other decoder is set to detect the high-frequency portion of the tone. If the proper key is pressed at the calling telephone, the digital output from both decoders will be logic 0. A CPU interprets this logic configuration and initiates an automatic playback cycle.

Notice that each tone decoder is set by using an adjustable resistor. If either of these resistors are adjusted carelessly, the desired tone might not be decoded properly, or an unwanted tone might be decoded instead. If the adjustable resistors are replaced with a switchable resistance network, it is very convenient to select various combinations of resistance that correspond to different telephone keys. This concept is often used in one form or another as a user-selectable security code.

Another remote control approach is illustrated in Fig. 8-7. The signal from a seized telephone line is applied to a fully-integrated DTMF decoder IC that contains all of the amplifier, the filter, and the digital conversion circuitry needed to produce a set of digital outputs. This set of digital outputs represents the binary equivalent of the selected digit. In this way, the answering machine can identify any dial pad keys (instead of only one key). A CPU interprets these digital codes and responds accordingly. For example, a caller pressing the numbers 2, 6, and 3 could start a message playback sequence, while the numbers 7 and 9 might cause the machine to

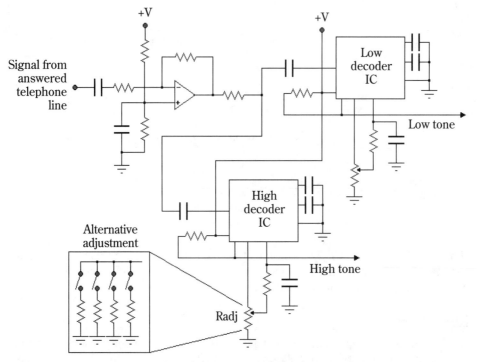

8-6   A remote control circuit.

rewind the ICM tape. A DTMF decoder allows a much broader range of remote control than the circuit shown in Fig. 8-6. DTMF decoder ICs have made a high level of remote control possible with virtually no increase in cost for the consumer.

Speech network

Although a speech network is not considered to be a large portion of an answering machine, it is certainly one of most critical portions. The signal from a seized telephone line is applied to the speech network which converts the two-wire tip and ring circuit to a four-wire circuit (two wires to transmit, and two wires to receive). In principle, the speech network serves the same purpose in an answering machine as it does in a telephone. However, an answering machine's speech network also includes switching circuitry that allows signals to be diverted throughout the answering machine as needed.

For example, a speech network must interface the playback signal of your OGM to the telephone line. A portion of that OGM is often fed to a speaker so you can hear the OGM and know that the machine is working. When the ICM cycle begins, received speech must be channeled to the ICM play/record head assembly, as well as to a speaker for monitoring any incoming messages.

The play/record system

The process of actually recording and replaying messages is handled by the play/record (P/R) amplifier IC as shown in Fig. 8-8. The P/R amplifier is under di-

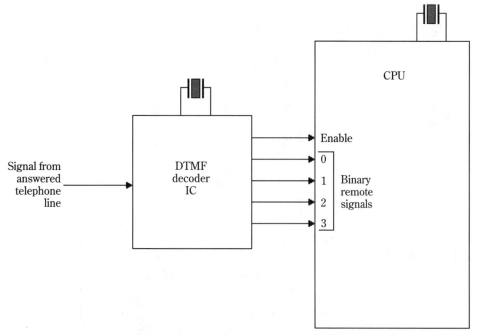

8-7 An integrated remote control circuit.

rect control of a CPU that determines whether the amplifier is handling an incoming or outgoing message, and whether the machine is in a play or record mode. The P/R amplifier returns its operating status (busy) and a logic signal whenever a message beep tone is detected (beep detect). Beep detection is important because it allows an answering machine to consistently find the beginning and end of messages.

When a new OGM is to be recorded, the CPU configures its control signals so that the OGM and record modes are selected, using the microphone signal source. If you are recording a new OGM by remote control, the machine will select the line as its signal source. The CPU engages the OGM solenoid and starts the tape drive motor. The erase head portion of the OGM play/record head assembly is activated to erase any previous message, and the new OGM is recorded to tape through the OGM play/record head. When the record cycle is complete, the machine advances the OGM tape until the beginning of the OGM is detected. The OGM solenoid is then released, and the tape drive motor stops.

When an ICM must be recorded, the CPU configures its control signals, so that the ICM and record modes are selected, using the line signal source. If you wish to leave a memo message on the ICM tape, the machine selects the microphone signal source. The CPU engages the ICM solenoid and starts the tape drive motor. The erase head portion of the ICM play/record head assembly is then activated to erase any previous message, and the new ICM is recorded to tape through the ICM play/record head. When the ICM recording is complete, the answering machine inserts a message end tone (a beep). The ICM solenoid is then released and the tape drive motor stops.

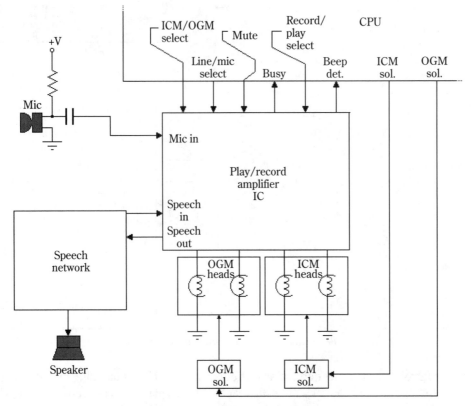

8-8 A play/record system (dual-tape).

During ICM playback, the CPU selects the ICM and play modes, using the line or monitor as its signal destination. The playback signal is also supplied to the monitor output which is coupled to the machine's speaker. The tape drive motor is engaged in reverse to rewind the ICM tape to its beginning. The ICM reel sensor stops when the ICM tape stops, so the CPU knows when the tape end has been reached. At that point, the ICM solenoid is engaged and the tape drive motor moves the ICM tape in the forward direction. The erase head remains unused during normal playback. Each message is separated by an audible beep. Every time a beep is encountered, the answering machine adds a count. When the count equals the number of recorded messages, the answering machine knows that all recorded messages have been played.

After playback, the ICM solenoid is disengaged, and the tape motor stops. This sequence leaves all recorded messages intact for future playback. If you wish to reuse a tape, you can rewind the ICM tape to the beginning.

Thus far, only dual-tape systems have been discussed, but single-tape answering machines are very similar in most respects. Instead of separate ICM and OGM head assemblies (each consisting of a play/record and erase head), a single-head set is used for both ICM and OGM operations. The single-tape approach means that your

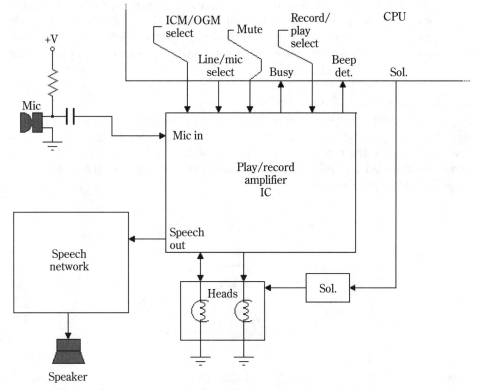

8-9   A play/record system (single-tape).

OGM is recorded on the same tape that holds your ICMs, as shown in Fig. 8-9. Such an approach often results in a smaller and less expensive answering machine, but the operating cycle is somewhat more involved.

Most of the control signals are identical to those signals found in a dual-tape system—it is still necessary for the answering machine to decide between ICM or OGM operation, the line or microphone/monitor signal, and the play/record mode. The actual operating cycles of single- and dual-tape answering machines are also very similar but only one set of magnetic heads is brought into contact with the tape.

The OGM plays first, then the tape advances to the nearest unused portion of tape before an ICM can begin recording. When the ICM is done, a beep is inserted to mark the message's end, then the tape is rewound to locate the OGM in preparation for another message. The OGM can be separated from ICMs using one of three methods: a different beep tone, a sequence of beep tones, or the physical location of the tape (OGMs are almost always located at the very beginning of the tape).

Messages are not erased during playback, so it does not take long for them to accumulate. It then becomes necessary to reset the tape so old ICMs can be overwritten. For dual-tape systems, this operation is as simple as rewinding the ICM tape to its beginning. On a single-tape system, however, the OGM is usually located at the tape's beginning. As a result, the single-tape reset typically rewinds the tape to its beginning, forwards the tape past the OGM, then fast-forwards the tape to its end with the erase head activated to eliminate any existing messages and beep tones. Once erased, the tape is again rewound to its beginning to prepare the OGM.

The motor drive system

Of course, the motor is at the heart of all tape-based answering machine operations—it provides the physical force that actually moves the tape(s). Notice from the diagram of Fig. 8-10 that a typical motor drive system also uses a series of discrete transistors (labeled Q1 through Q6) to channel current through the motor as required by CPU control signals.

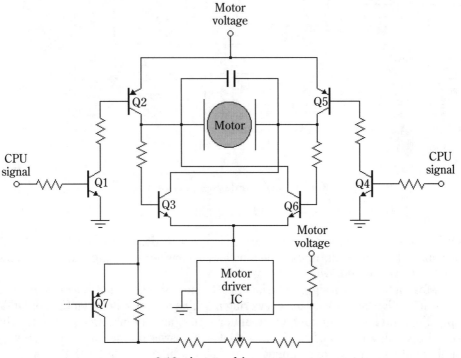

8-10   A motor driver system.

When the motor is intended to turn counterclockwise, a logic signal from the CPU activates Q1 which, in turn, activates Q2 and Q3 (Q4 through Q6 remain off). Motor voltage causes current to flow through Q2, through the motor, then through Q3 to a motor driver IC. If clockwise rotation is desired, the CPU sends a logic signal to Q4. The output from Q4 also activates Q5 and Q6. Transistors Q1 through Q3 remain off—only clockwise or counterclockwise rotation can be selected by the CPU. Motor current flows through Q5, through the motor, then through Q6 to a motor driver IC. The motor driver IC is used to regulate motor current.

Keep in mind that not all answering machine motor drive systems use discrete transistor networks. The transistor arrangement shown in Fig. 8-10 could easily be integrated onto an IC chip.

Controls and displays

Even the simplest answering machine requires some form of indicators and user controls. The exact type, quantity, and configuration of indicators and controls will depend on the particular manufacturer and model of answering machine. As a minimum, your machine must provide controls for the tape system (i.e., playback, fast-forward, rewind, and stop). A minimum display includes some form of message counter, along with an indicator to display the machine's status. Figure 8-11 illustrates such a simplified display system driven directly by a CPU.

In order to show the exact number of waiting messages, the message LED is made to blink on and off over a set period of time. For example, if three messages are waiting, the LED might blink three times quickly, remain dark for 1 or 2 seconds, then repeat the blinking pattern. Although such a design is electrically simple and very inexpensive to implement, it is often quite inadequate for displaying more than six or seven messages.

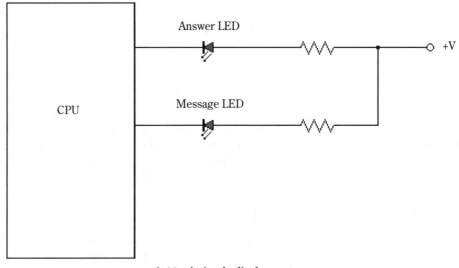

8-11 A simple display system.

A much more user-friendly message counter is shown in Fig. 8-12. A CPU generates data that drives an array of seven-segment displays. The display device itself might be LED or LCD in nature, but both would serve the same purpose. The CPU can drive the display directly, or its data can be amplified by a driver IC. In addition to the message counter, the CPU can operate other discrete LEDs.

The machine's control panel is often configured in a fashion very similar to the diagram shown in Fig. 8-13. Voltage is applied to each column of switches. When a switch is closed, current flows through a current-protecting diode, then through the

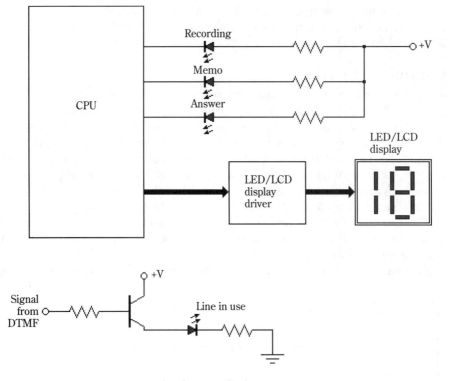

8-12   An alternate display system.

selected switch. Voltage, now present at the switch, causes a signal change at the row and column pins of the respective switch. The CPU detects any change in row and column signals and determines which key has been pressed. The CPU is usually capable of processing switch signals directly.

Tape system mechanics

No discussion on answering machines would be complete without at least an overview of the tape system mechanics. The purpose of a mechanical system is deceptively simple—to move magnetic tape back and forth at a constant speed, tension, and height relative to a stationary magnetic head assembly. If either the tape's speed, tension, or height is incorrect, speech will not play or record properly. A diagram of a typical tape system is shown in Fig. 8-14.

At the center of the system is, of course, the main motor. To save weight and cut down on space requirements, the motor is usually selected as a small, low-torque (force), high-speed device. The motor's rotating shaft is coupled by a belt to a large flywheel that reduces the rotational speed of the motor—while increasing the motor torque needed to drive the OGM or ICM tapes.

The series of gears and linkages under a tape chassis is used only to transfer the motor's force and move a tape. You should never apply lubricants to any portion of a

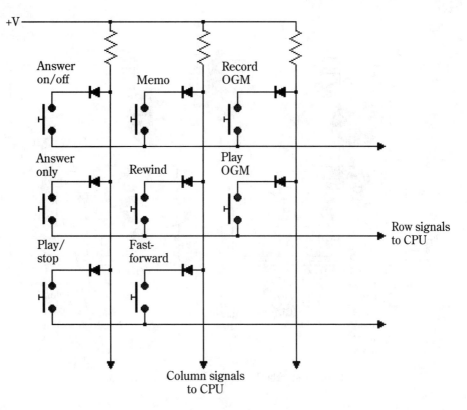

8-13 A control panel.

tape mechanism. Most of the mechanical parts are very small and constructed of plastics or plastic composites, so friction is almost negligible anyway. Lubricants would only serve to introduce physical drag in such a delicate assembly.

When a head assembly is engaged to a tape, the tape must run evenly and under constant tension. To stabilize tape tension, a *pinch roller* engages to press the tape against a series of *idle rollers*. Each roller's surface is made from rubber or another soft elastomer that can come in contact with a tape's surface without damaging it.

However, when contact is made, some of the tape's magnetic oxides do rub off onto the rollers. Not only does this eventually degrade the tape, but the buildup of magnetic materials causes distortion to any information contained on the tape. A roller area should never be lubricated, but rollers should be cleaned periodically with a cotton swab dipped lightly in denatured alcohol or other non-solvent cleaner designed specifically for magnetic tape systems. Tape heads should also be thoroughly cleaned the same way to remove any buildup of magnetic oxides.

The tape head assembly is not intended for adjustment without the proper test equipment—but there are two common adjustments that you should be familiar with (so you can at least avoid them). These adjustments are pointed out in Fig. 8-14. A tape speed adjustment does just what the name implies—overall tape speed can be altered

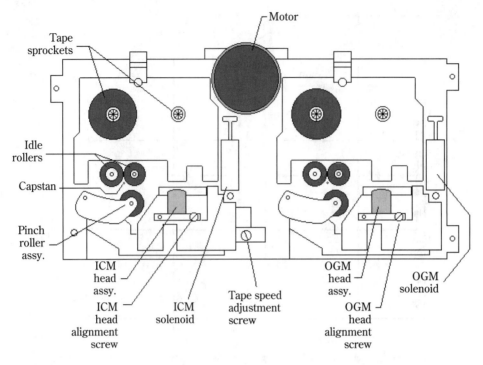

8-14 A standard dual-tape mechanical system.

within narrow limits to optimize tape sound and ensure that each machine runs at exactly the same speed. A change to the tape speed adjustment will have little impact on the operation of your particular machine, but tapes recorded on such a machine will sound weird if played back on other machines or cassette players.

A second adjustment you should know about is the head assembly adjustment. Each head assembly has its own adjustment. Unlike the tape speed adjustments, head adjustments have a profound impact on the quality of sound, because the head's actual position relative to the tape surface is affected. Do not attempt to adjust a head's position unless you have the specific test equipment and procedures to guarantee the proper alignment. If you must make an adjustment anyway, be certain to clearly mark the original position so you can return to it later if you get in trouble.

Non-tape systems

The rapid advances of digital electronics are starting to be felt in answering machine applications. Cassette tapes are gradually being replaced by integrated circuits used to store outgoing and incoming messages. In essence, analog speech signals are digitized (converted to corresponding digital information) at high rates, then placed in solid-state memory. During playback, stored digital information is recalled at the same rate at which it was recorded, reconverted into corresponding analog levels, then filtered to approximate a smoothly varying analog signal that resembles the speaker's voice. The recorded signal can then be supplied to the speech network for transmission over a telephone line—or to the answering machine's speaker.

To reduce the amount of raw memory required to hold messages, a digital signal processor (DSP) IC capable of compressing speech data can be added to the circuit. As speech is recorded, information is processed through an algorithm that yields fewer pieces of data to represent the same amount of speech. When the data is played back, it is processed through a reverse-algorithm to insert the appropriate pieces of information that were eliminated during compression. This portion of the chapter briefly examines the basic concepts of digitized speech, and looks at simple answering machine applications.

Analog-to-digital conversion

The first step in the conversion of speech into digital information is to transfer the acoustical energy of speech into electrical energy, as shown in Fig. 8-15. When recording a memo or OGM, such transfer is accomplished by a microphone. If speech is being digitized from the telephone network, the signal is already in its electrical form. The analog signal must then be amplified to a level that is adequate for conversion. A simple operational amplifier circuit is usually sufficient.

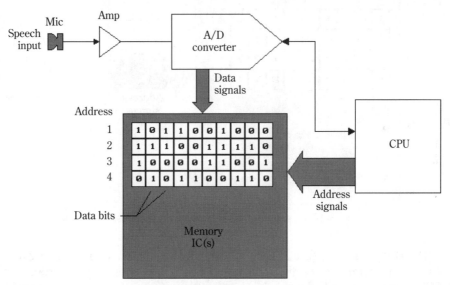

8-15 A simple digitizing process.

At that point, the analog signal is applied to an *analog-to-digital converter* (ADC). An ADC takes samples of the analog signal at very high rates and converts those samples into corresponding digital words—typically at the rate of 5000 to 8000 conversions-per-second. The CPU monitors and regulates the conversion process, then stores the data generated by the ADC into memory locations. Ordinary tapeless answering machines typically store up to ten messages—each message can be up to 2 minutes long.

Digital-to-analog conversion

When an OGM or an ICM is played back, digital data must be recalled from memory and reconstructed back into analog information before it can be heard through a speaker or connected to a speech network. The CPU monitors and regulates the reconstruction by cycling through each memory location associated with a particular message, then passing data from each location to a *digital-to-analog converter* (DAC) as illustrated in Fig. 8-16.

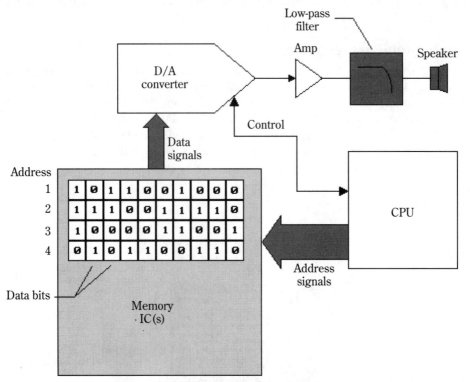

8-16 A simple voice reconstruction process.

The DAC translates data at each memory location into a corresponding analog voltage. Data is processed at the same rate at which it was originally recorded, so the DAC's output should yield a rough approximation of the original signal. Once in its analog form, the reconstructed speech signal is sent through a low-pass filter to attenuate any unwanted high-frequency signals that might have been produced during reconstruction. After filtering, the signal should be a fair approximation of the speaker's original voice.

Applying digitized voice

Figure 8-17 shows a block diagram of a typical answering machine. It employs a digitized speech network to handle both the OGM and ICMs. As you can see, most of

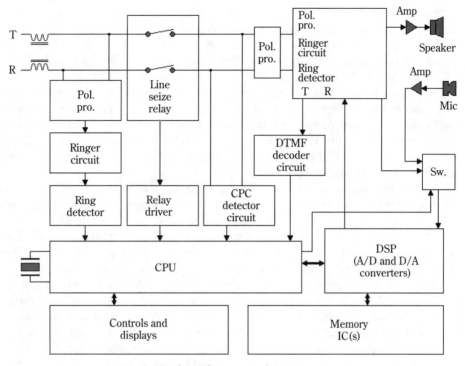

8-17   A tapeless answering system.

the functional areas are exactly the same as with the tape-driven system of Fig. 8-2—except for the data conversion and memory circuits. Notice that all of the tape drive hardware (motor, motor-driver circuits, tape reel sensors, solenoid mechanisms, recording heads, etc.) has been removed from the machine—only integrated circuits are employed, so there are no moving parts. The result is a smaller, lighter, often more reliable package than is available with standard mechanics.

Troubleshooting answering machines

Now that you understand the overall design and functional areas of both tape and non-tape answering machines, you can move into the troubleshooting and repair techniques. As with electronic telephones, you will undoubtedly encounter a variety of PC boards and mechanical assemblies that can differ substantially with each model and manufacturer. However, each answering machine offers the same general functional areas as illustrated in Figs. 8-2 and 8-16.

Symptom 1   Message tape does not move at all (ICM or OGM). Check the tape before disassembling any portion of your answering machine. If the take-up spindle is moving, but the source spindle is not, check to make sure the tape is not broken. If both spindles appear to move erratically, see that the tape is not tangled (or eaten) by the pinch roller assembly. Check the tape wheels by hand to see that the tape moves freely from wheel to wheel. If the tape is jammed internally, discard the tape and replace it. If the cassette tape is intact, you can play it in another cassette player to make sure that the tape is working properly.

If the tape itself is not the problem, turn off all power to the machine, inspect the overall mechanical tape assembly, and check the tension of the flywheel belt. The belt should be seated properly and be snug—not loose. Replace any belt that is broken, excessively worn, or loose. A faulty belt will not transfer motor force to the flywheel, so none of the tape drive mechanics will operate properly. You should use an exact replacement part to ensure that the belt applies tension and friction within the machine's working limits. Belts that are too loose or made of the wrong material can slip. Belts that are too tight can snap—or apply so much friction to the mechanism that it puts an excessive load on the motor.

Inspect the defective tape mechanism closely, and check for any gears, springs, or linkages that might be jammed, bent, or broken. In a dual-tape system, you can compare mechanical actions between a working and suspect drive mechanism to identify possible trouble spots. Be very careful when working with tape drive mechanisms. Many parts are small and extremely delicate. A large number of parts are made from plastic. You can easily cause additional damage by careless handling or rough prodding.

If the motor does not turn at all, but the mechanics appear to be operational, the trouble probably lies in one or more of the motor's power transistors, or in the motor-driver IC. Remove all power from the answering machine, disconnect the motor's leads, and use your multimeter to measure the resistance of each motor winding. Compare your measured readings to any ratings marked on the motor. If no markings are available, a good rule of thumb is 5 to 50 Ω. If you read a very high resistance (or infinity), the respective winding is probably open-circuited. If you read a very low resistance (or a short circuit), the winding is probably shorted. In either case, the motor should be replaced.

With the motor disconnected, reapply power briefly, engage a tape action (i.e., play, rewind, fast-forward, etc.) and use your multimeter to measure the motor voltage across the open motor terminals, as shown in Fig. 8-18. You should read a strong, steady voltage of between 6 to 12 Vdc or more. If motor voltage is absent, the motor is not receiving power when an action must be taken. Remove power and check the motor-driver transistors and IC. If motor voltage appears to be present, connect the motor and repeat your measurement. A short-circuited motor will substantially reduce the amount of available motor voltage. Replace the suspect motor.

Remove power from the machine and check the motor-driver transistors as discussed in chapter 4. Replace any transistors that appear to be open- or short-circuited. Be certain to replace defective transistors with ones of the same type and ratings.

Current that powers the motor is regulated through a motor-driver IC. If the driver IC fails, current can be cut off, preventing the motor from operating. If everything else is functioning properly, try replacing the motor-driver IC. Finally, try replacing the answering machine's CPU. The CPU might not be generating the proper control signals needed by the motor-driver.

Symptom 2 No audio is produced during tape play (ICM or OGM). First, make sure that something has been recorded on the tape to begin with. If the ICM or OGM failed to record, there will be no message to play. You can try playing the tape in an ordinary cassette player. If you hear a new message on the tape, the answering ma-

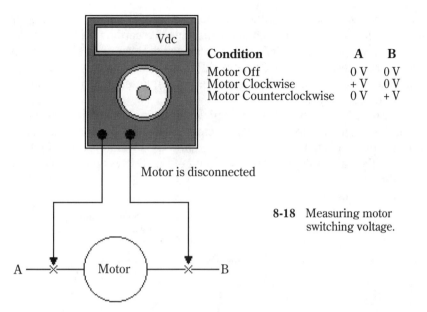

Condition	A	B
Motor Off	0 V	0 V
Motor Clockwise	+ V	0 V
Motor Counterclockwise	0 V	+ V

Motor is disconnected

8-18 Measuring motor switching voltage.

chine playback system is probably defective. If a cassette player does not reveal any new messages, then the message was not recorded, or the tape could be defective.

During a playback operation, signals picked up by the play/record head are conditioned and amplified by the play/record amplifier IC, then sent along to the speech network or speaker. If you have an oscilloscope with a transformer-coupled probe, you can probably watch voice signals from the magnetic head into the P/R amplifier, and from the P/R amplifier into the speech network. Keep in mind that voice signals picked up by a magnetic head are very small, so your oscilloscope's sensitivity will have to be set very high (in the microvolts-per-division region). Apply power to the machine and start a playback cycle. Make sure the corresponding head assembly engages to the tape properly, and measure the P/R amplifier's input.

If there is no input to the P/R amplifier, the magnetic head assembly is probably defective. Replace the magnetic head assembly. If you find an output from the P/R head, but no output from the P/R amplifier (you might have to reduce your oscilloscope's voltage sensitivity to obtain a clear trace of the P/R amplifier's output), the P/R amplifier IC is probably defective. Replace the P/R amplifier IC.

If the P/R amplifier produces an output, but nothing is audible on the machine's speaker, the speech network IC is probably defective. Replace the speech network IC. Trace speech from the speech network IC to the speaker. Some answering machines use solid-state switching to cut the speaker in and out as needed. Such a switch might have failed. Also, the speaker amplifier IC—or even the speaker itself— might be defective. If the message is not being heard on the telephone line, trace the signal from the speech network to the machine's output. The point at which the speech signal disappears is probably the point of your circuit failure.

If you do not have an oscilloscope, there is a general order of procedures that you might like to try. In many cases, part or all of the P/R amplifier has failed. Try replacing the P/R amplifier IC and retest the machine. If that does not correct the problem, re-place the P/R magnetic head itself. If the machine is recording messages, it is unlikely that the speech network has failed, but there could be an open circuit that prevents speech from entering the speech network IC. Try replacing the speech network IC. Finally, check the speaker and its associated wiring to make sure that the speaker is intact and connected properly.

Symptom 3   Machine does not pick up on incoming ring. Ac ringing volt-age must be converted to logic pulses before a CPU can count them. If any portion of the ring detection network is interrupted, the CPU will never pick up the ringing line and initiate a message cycle. The circuit of Fig. 8-19 is an example of a ring de-tection circuit.

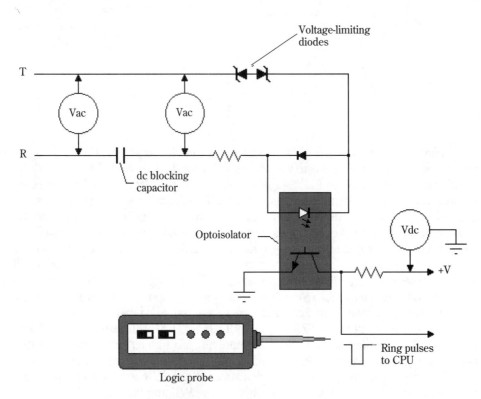

8-19   A partial schematic of a ring detection circuit.

You must first determine if ringing voltage is reaching the detection circuit. Use your multimeter to measure the ac ringing voltage (typically 90 Vac) across tip and ring. If ring voltage is present during normal ringing, measure ring voltage on the de-tector side of the dc blocking capacitor. If ring voltage then disappears, replace the defective dc blocking capacitor. Make sure that all power and telephone signals are disconnected from the machine before attempting to replace a component.

With power disconnected, use your multimeter to check the zener diodes in se-ries with the tip and ring path. Zener diodes are typically low-voltage devices (often rated around 24 volts) that prevent current from flowing in the detection circuit until ring voltage exceeds the zener voltage. In this way, zener action prevents false triggering of the ring detector. Chapter 4 supplies the procedures for checking semiconductor diodes. Make sure that all circuit power is off before attempting to make diode checks.

Use your logic probe or oscilloscope to measure the presence of ring pulses sent along to the CPU. During the audible portion of a ring, the optoisolator's output is configured to supply logic pulses to the CPU, so your logic probe should measure a clock or pulse reading. If your logic probe displays a steady logic 0 or logic 1 during the audible portion of the ring, either the optoisolator is defective, or the circuit's logic voltage is missing. If the machine's logic voltage (about +5 Vdc) is present, replace the optoisolator. A constant logic 1 during the audible portion of a ring suggests that the optoisolator is defective.

Finally, your last option is to replace the answering machine's CPU. If logic pulses are being generated by the ring detector as expected and the CPU does not respond, the CPU might well be defective.

Symptom 4   No beep tone is generated (or played back) at the end of messages. When an answering machine has finished recording an ICM, the CPU triggers a brief, fixed-frequency tone (called a beep) that is added to the tape to separate messages from one another. Each time a beep is added, the answering machine's message counter increments by one. During ICM playback, the CPU detects each beep and decrements the message counter by one. Using this method, an answering machine will replay only the exact number of messages that it recorded since the last time it was played back or reset—regardless of how many miscellaneous messages are scattered throughout the tape.

If no beep is generated or recorded to tape, the CPU won't be able to count calls or keep track of where messages begin or end. In most cases, the failure lies in the CPU, or part of the associated circuitry that carries the tone to or from the tape. Remove power from the machine and try replacing the CPU.

The play/record amplifier IC is responsible for carrying signals to and from the tape. A defect in the P/R amplifier will also cause a similar problem. If a new CPU does not correct the problem, replace the P/R amplifier IC.

Symptom 5   A tape does not rewind. Begin your inspection with the tape drive mechanism. Remove all power before disassembling the machine, then check for any loose, bent, or broken gears or linkages that might prevent the motor's force from rewinding the tape. For dual-tape systems, you can often compare the OGM and ICM mechanisms to help find defective parts.

If the mechanism itself appears to be intact, and the motor does not operate in the reverse direction, you should check the motor's driver transistor array. Transistors are often used to switch voltage polarity under control of a CPU. By reversing polarity, a motor's direction of rotation can be reversed. As shown in Fig. 8-10, transistors Q1, Q2, and Q3 drive the motor in one direction, while transistors Q4, Q5, and Q6 drive the motor in the opposite direction. Of course, both sets of transistors cannot be on simultaneously.

Use your multimeter to measure the dc motor voltage, then measure across the motor while the machine is in the play mode. You should read a motor voltage of about 6 or 12 Vdc (voltages might read negative, depending on the orientation of your meter leads). Set the machine to rewind. Your dc voltage reading should reverse and result in a reversed motor direction. If motor voltage drops to 0 Vdc while the motor is stopped, then remains at 0 Vdc when the machine is placed into the reverse mode, remove power from the machine and use your multimeter to check each driver transistor. Replace any defective driver transistors and retest the machine.

If all driver transistors check out properly, the defect probably lies with the CPU. If the CPU does not supply the necessary logic pulses to operate the transistor array, the motor will simply not be activated. Reapply power to the machine and measure logic signals that are driving the transistor array. In the play mode, one signal should be a logic 1 and another signal should be logic 0. To reverse the motor, these logic signals should reverse. If one of these logic control signals does not change, the CPU is defective and should be replaced. If you find the logic control signals to be intact, take another look at your transistor driver array and motor connections.

Symptom 6   OGM is heard through the speaker, but not transmitted to the telephone line. If an OGM is heard at all, it usually indicates that the playback portion of the OGM P/R head and P/R amplifier IC are working properly. The task becomes one of finding the place where the played signal disappears. In most cases, the speech network IC (or at least part of it) is defective. You can verify this by measuring the voice signals into and out of the speech network with an oscilloscope using a transformer-coupled probe. If you find the OGM voice signal entering the speech network, but no signal is leaving for the telephone line, the speech network IC is probably defective. Turn off all machine power before replacing the suspect IC.

If speech is present at the speech network output, you must carefully inspect the circuit for any PC board breaks or component failures (such as open transistors) that might be interrupting the speech path to the telephone line. If you are fortunate enough to have access to an oscilloscope with a transformer-coupled probe, you can probe along the PC board toward tip and ring until the speech signal disappears. That point will probably be the source of your problem.

Symptom 7   New ICMs are mixed together with old ICMs (old ICMs are not being erased). When an ICM cycle engages, the erase head on the ICM P/R head assembly is activated to erase any previous messages before the P/R head records a new message. In many instances, the erase head or its associated wiring has failed and become an open circuit. Without an erase function, new messages will simply be superimposed over old messages. Replace the erase head and retest the system.

If the erase head itself is not at fault, the erase head driver circuit is probably defective—the erase head is just not being energized. The erase head driver circuit is typically contained in the P/R amplifier IC, so remove all power from the machine and replace the suspect P/R amplifier IC.

Symptom 8   Machine disconnects the telephone line in only a few seconds after an ICM starts. A working answering machine should record an ICM until one of four conditions arises:

1. The fixed message time limit is reached.
2. The caller hangs up.

3. The ICM tape runs out.
4. The caller stops talking for more than a few seconds.

The circuit that detects the presence of voice is called the *vox* circuit, and is usually located in the speech network. As long as a caller speaks, the vox produces a logic signal that a CPU can recognize and use to continue recording.

If the vox fails to produce the appropriate output, the CPU will be fooled into thinking that the caller has stopped talking. Without a vox signal, the CPU will shut down the ICM cycle to conserve tape. Remove all power from the machine and replace the vox circuit IC. If the vox circuit uses discrete components, use your multimeter to check each component. Replace any component that appears defective.

Symptom 9   A new OGM will not record. You should be able to record new OGMs at any time. Many answering machines with full DTMF remote control even allow you to record new OGMs remotely. When you engage a new OGM record cycle, the OGM tape should begin to move, and you can speak into the machine's built-in microphone (or the calling telephone if recording by remote control). If your new message is not being committed to tape, however, there are several things you can check.

Begin by checking the microphone and its associated wiring. A defective microphone will not pick up the full range of sound that it is intended to—if it picks up anything at all. Secure any loose wires or connectors. If a new OGM records remotely, but not from the local microphone, try replacing the microphone.

If no OGM records at all, the OGM P/R head might be defective. Check the OGM P/R head wiring very carefully. Try replacing the OGM P/R head. If you have access to an oscilloscope with a transformer-coupled probe, you can often use a sensitive input setting to detect the OGM signal at the head. If an OGM signal is present during recording—but no speech is recorded to tape, replace the defective P/R head. If OGMs still cannot be recorded, the associated P/R amplifier IC is probably defective and should be replaced.

Symptom 10   No control provided by remote. In most answering machines, some type of DTMF decoder arrangement is used to decode DTMF tones from a calling telephone and provide logic signals to the CPU for processing. When DTMF remote control fails, the DTMF tone decoder(s) are usually at fault. Replace the DTMF tone decoder IC and retest the machine.

Older remote control circuits are slightly more complex because separate tone decoder ICs are used to isolate both the high and low frequency portions of a DTMF tone. Each IC must be tuned precisely for the desired frequency, or else the wrong frequency pair will be detected. You can use a logic probe to measure the logic output of each tone decoder.

Press the desired key on a local telephone and verify that the logic output changes as expected. If an undesired key causes a logic change, the tone decoder is maladjusted and will have to be realigned. If no key produces the desired output, the tone decoders are either hideously out of alignment, or one or both tone decoder ICs are defective.

In some circuits, an amplifier might be needed to condition the DTMF signal before it is distributed to the tone decoders. If the amplifier is defective, all DTMF signals might be cut off from the tone decoders. Try replacing the DTMF amplifier IC. When your tone decoder circuit or DTMF decoder IC is working properly, the trouble might exist in your CPU. Try replacing the CPU.

Symptom 11 The machine does not release the telephone line when the caller hangs up. When a caller hangs up, your local CO might produce a short calling party control (CPC) pulse that the answering machine can detect and use to disconnect itself from the telephone line. One common problem is related to the CPC pulse width itself. CPC pulses produced by most COs are either 10 ms or 350 ms. Most answering machines can be set to detect either short or long CPC pulses (they can be switched to match the pulse provided by your particular CO).

However, an incorrect CPC setting on your machine can easily prevent the CPC disconnect from working. For example, suppose that your machine is set to detect long CPC pulses while your CO produces short CPC pulses. Your machine will expect to find 350-ms pulses, but it will only receive 10-ms pulses. The difference is so great that the CO's CPC pulse will be ignored. Check your CPC setting and retest the machine.

If the machine's CPC switch is set properly, use your multimeter to check the voltage output from the CPC circuit diode bridge. An open diode can prevent the circuit from recognizing an offhook condition. If voltage output is absent when the machine is in its offhook state, remove all power and use your multimeter to check each bridge diode as discussed in chapter 4.

CPC pulses are so short and irregular that they are very difficult to see reliably—even with an oscilloscope—unless the scope is triggered properly. If the CPC bridge rectifier checks correctly, just replace the CPC optoisolator outright and retest the machine.

The logic output from a CPC circuit can be amplified using a discrete transistor or transistor network. Remove power from the machine and use your multimeter to check each discrete driving transistor. Replace any driver transistor that appears defective. Refer to the procedures in chapter 4 for information on transistor checking.

9
Cordless telephones

CORDLESS TELEPHONES, SUCH AS THE UNIT SHOWN IN FIG. 9-1, REPRESENT A unique marriage of FM radio and electronic telephone technologies that allows the major functions of an electronic telephone to be carried significant distances from the original telephone jack without the impediment of a physical cable connection.

As with telephone equipment in general, cordless telephones are no longer the heavy, awkward, battery-hungry devices of even a decade ago. Designers have been quick to capitalize on the use of sophisticated ICs to improve cordless performance, range, and reliability—while reducing size, weight, and power requirements. Such design improvements have been instrumental in bringing cordless telephones into everyday home and business use. This chapter examines the operations and troubleshooting of typical cordless telephones.

Anatomy of a cordless telephone

In reality, a cordless telephone is not one device, but two separate devices, as shown in Fig. 9-2. These two devices are known as the *base unit* and the *portable unit*. Both units are connected by means of an AM (amplitude modulated) or FM (frequency modulated) radio link.

A base unit accepts regular ac power from standard receptacles. It powers not only the base, but a battery charger for the portable unit as well.

The base contains the telephone electronics needed to interface to the local loop, along with some control circuitry and a complete radio transmitter and receiver. The control circuitry includes such things as a ring detector and ringer, loop seize relay, DTMF dial pad or tone generator IC.

Courtesy of Radio Shack.

9-1   A cordless telephone with number memory and redial features.

The portable unit is powered by a rechargeable battery pack and contains (as a minimum) a complete radio transmitter and receiver, a DTMF dial pad, a microphone, and a receiver element. Both the base and portable units are controlled by independent CPUs.

In order to establish full-duplex operation in a cordless telephone (speaking and listening simultaneously), two distinctly different frequencies are required. One frequency carries signals from the base unit to the portable unit, while another frequency carries signals from the portable unit to the base unit. These full-duplex frequencies are arranged in carefully selected pairs to assure that there will be no interference between transmitted or received signals.

Older cordless telephones used a single set of frequencies. While this system proved to be effective under most circumstances, radio interference from external

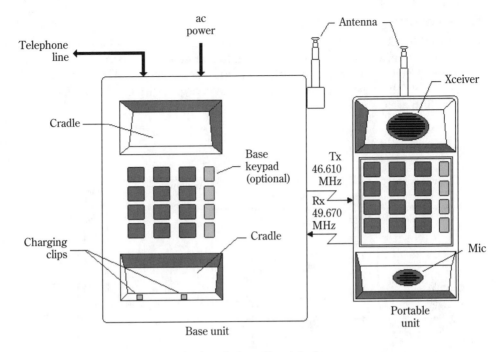

9-2 A typical cordless telephone.

sources or other nearby telephones could not be compensated for. Today's cordless telephone sets offer a selection of frequencies that can be switched to avoid unexpected interference with other pieces of electronic equipment. Having a variety of frequency sets is not a foolproof answer to every possible interference problem, but it does allow cordless telephones to be used under a wider range of circumstances. Table 9-1 shows a typical selection of cordless telephone frequency pairs.

**Table 9-1. A selection of
cordless telephone frequency pairs.**

Channel	Xmit (Portable—MHz)	Xmit (Base—MHz)
1	49.670	46.610
2	49.845	46.630
3	49.860	46.670
4	49.770	46.710
5	49.875	46.730
6	49.830	46.770
7	49.890	46.830
8	49.930	46.870
9	49.990	46.930
10	49.970	46.970

The base unit

A simplified block diagram of a typical base unit is shown in Fig. 9-3. As you can see from the diagram, a base unit can be divided into four general areas: a receive circuit, a transmit circuit, the telephone interface circuit (speech network), and the control (CPU) circuit. The base unit's power supply and charging circuits are often considered as a fifth area.

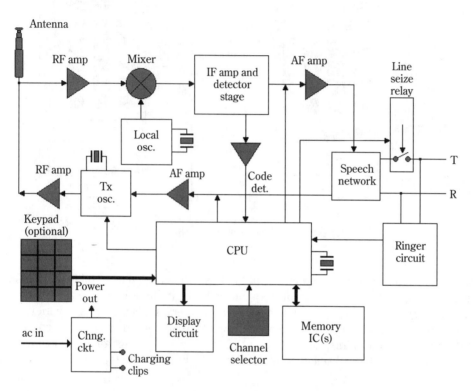

9-3   A typical base unit.

Signals received from the portable unit are isolated by the base unit's antenna and fed into an RF (radio frequency) amplifier circuit. An RF amplifier boosts the very faint speech signals received from the portable unit. A typical RF signal contains three things:

1. An RF carrier (for example, a sine wave of 49.670 MHz).
2. Speech signals (ranging from 0 to about 4 kHz).
3. Control signals that coordinate operation between the base and portable units.

One good example of a control is the hook switch control (sometimes known as talk). When the talk button is pressed on the portable unit, a control signal is picked up by the base unit, interpreted by the CPU, and used to activate the line seize relay.

Once the relay is engaged at the base, dial tone from the local central office is transmitted to the portable unit where it is audible.

In order to obtain speech and control signals from a received RF signal, the RF portion of the signal (the carrier) must be removed. The process of extracting useful signals from RF is known as *demodulation*. Demodulation uses a mixer circuit to combine the incoming RF signal with the output of an on-board (or local) oscillator. The local oscillator runs at a frequency somewhere between RF and AF (audio frequency) ranges, and that frequency is known as the IF (intermediate frequency). The mixer's output contains four combinations of frequencies, as well as a selection of unwanted harmonics but the signals of interest are the audio signal and carrier (now the IF signal)—all other unwanted signals can be filtered out

The desired IF signal is then demodulated further in the IF stage by detecting and separating the AF speech and filtering out the IF carrier. Control signals are channeled directly to the CPU in the control circuit, while speech signals are amplified and coupled to a hybrid transformer or speech network IC for transfer to a normal telephone line.

The process of speech and control signal transmission is a bit simpler. Speech signals coupled from a regular telephone line by a hybrid transformer or speech network IC are fed to an AF amplifier. Amplified speech (along with any control signals) is then converted to an RF signal (or *modulated*) by combining it with a transmitter (or *carrier*) oscillator. The modulated signal is then amplified by an RF amplifier, and supplied to the base unit's antenna assembly.

A control circuit is the heart of a cordless telephone. In a base unit, the CPU coordinates the process of transmission and reception, generates control signals that are sent to the portable unit, interprets any control signals received from the portable unit, handles ring detection, generates desired pulse or tone dialing signals, and manages the local loop through the telephone interface. The CPU might also use one or more memory ICs to hold fixed program instructions and data.

If your particular base unit offers a built-in keypad, the keypad's row and column data is normally interfaced directly to the CPU. As you might suspect, a CPU failure cannot only disable a base unit, but render the associated portable unit useless as well.

Finally, the telephone interface (speech network circuit) handles the base unit's connection to the local loop. When the cordless telephone is activated by the portable unit, a line seize relay in the base unit closes and couples the telephone line to the speech network. Loop current is drawn from the local loop, and the central office will recognize that your telephone is offhook. Audible dial tone received from the central office is modulated and transmitted to the portable unit. Dialing can then begin from the portable unit.

The portable unit

As you can see from the simplified block diagram of Fig. 9-4, a typical portable unit generally contains the same basic working areas as does the base unit: a receive circuit, a transmit circuit, and a control (CPU) circuit. Many portable units contain a dial pad that is interfaced directly to the CPU. Dialing generates control signals that are

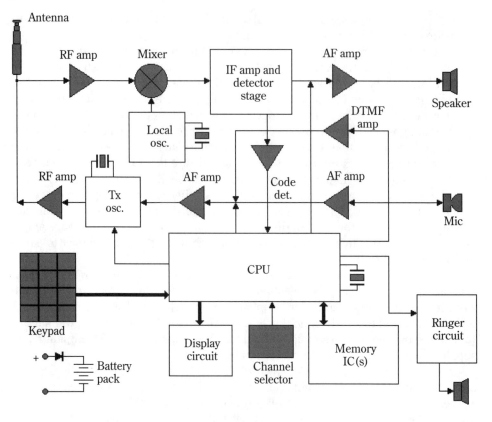

9-4 A portable unit.

passed along to the base unit, translated to corresponding DTMF tones, and coupled to the telephone line. A very simple charging circuit is also included to keep the portable unit's battery charged when it is idle in the base unit.

Signals received by the portable unit's antenna are fed into an RF amplifier that acts to boost the weak radio signals transmitted from the base unit. A typical RF signal contains radio carrier (46.610 MHz for example), speech (ranging from 0 to 4 KHz),and control signals from the base unit to coordinate the telephone's operation. In order to isolate speech and control signals from the RF combination, the carrier portion of the signal must be removed through very selective filtering. As with a base unit, the process of demodulation is accomplished using a mixer circuit to combine the complete RF signal with the output of a local oscillator. The local oscillator operates at an intermediate frequency (IF) that is somewhere between AF and RF ranges. The mixer's output contains four combinations of frequencies—plus unwanted harmonics—but the only signal of interest is the audio signal and carrier (now the IF). Unwanted frequency combinations and harmonics are removed by very sharp, selective filtering in the IF stage.

The IF carrier is then demodulated further by detecting and separating the AF speech or control signals while filtering away the IF carrier. Control signals are amplified and channeled directly to the CPU in the control circuit, and speech signals are delivered to an audio power amplifier and receiver element (often a speaker).

Transmission in a portable unit is accomplished very much like the approach used in a base unit. Voice signals from the transmitter element (usually an electret or electrodynamic microphone) and control signals from the CPU are combined and amplified by an AF amplifier. The amplified AF signal is modulated onto an RF carrier produced by a carrier oscillator, amplified by an RF power amplifier stage, then supplied to the portable unit's antenna.

A control circuit is the heart of a portable unit. The CPU coordinates transmission and reception, generates control signals and DTMF signals that are sent along to the base unit, and interprets control signals that are received from the base unit. When an incoming ring is detected by the base unit, the portable unit's CPU fires a ringer to alert you to the call. Finally, the portable unit of most cordless telephones holds a full keypad of dialing digits and control buttons. Key signals are interfaced directly to the portable unit's CPU. The control circuit might also contain some memory ICs to hold the CPU's program instructions and data.

Full duplex performance

After reviewing the block diagrams of Figs. 9-3 and 9-4, you might wonder how the transmitter circuit output and receiver circuit input can be connected to a single antenna at the same time and not have problems with feedback or other interference. Why doesn't the transmitted signal come back through the receiver? The answer to that question lies in two very important factors: the use of two different frequencies to transmit and receive, and use of extremely selective filtering in the RF stage of a receiver circuit

Consider the portable unit illustrated in Fig. 9-4. It uses a transmit carrier of 49.970 MHz and a receive carrier of 46.970 MHz. When your voice signal is modulated with a 49.970-MHz carrier and placed on the antenna, the signal is also made available to the receiver circuit's RF stage. However, the receiver circuit is tuned to accept only signals riding on a 46.970-MHz carrier—all other frequencies, including your transmitted signal, will be filtered away and ignored. At the same time, any signals that reach the portable unit's antenna with a 46.970-MHz carrier will be passed through to the RF stage and demodulated normally. In this way, transmission and reception can take place simultaneously.

Cordless troubles

Unfortunately, for all their conveniences, cordless telephones are not without their inherent disadvantages. Keep in mind that disadvantages are not necessarily defects or design flaws, but simply part of the device's nature. Cordless troubles can usually be grouped into four areas: battery problems, range limitations, radio interference, and radio privacy.

Battery problems

Portable units are powered by rechargeable battery packs, often consisting of one or more NiCad (nickel-cadmium) batteries. When the portable unit is off and idle on the base unit, a set of contacts between both units allows the battery pack to be charged from the base unit's power supply. Although this is a convenient and effective method of keeping a battery pack charged, NiCad batteries suffer from limitations that you should be aware of.

NiCad batteries have a rather low energy density as compared to other kinds of non-rechargeable batteries. Because their energy density is relatively low, NiCad batteries cannot provide high energy levels for a long period of time (without recharging). While the physical materials and construction of NiCad cells has improved, and better electronic circuitry has lessened the power demands on NiCad cells, you should not expect more than about 10 hours of use from the portable unit without recharging. Most telephone calls take much less than 10 hours, so this is rarely a problem.

NiCad batteries can also become troublesome when they are frequently discharged to some regular level, then recharged again. This can happen, for example, if you spend an average of 20 minutes per call many times each day, allowing the portable unit to recharge between calls. Such partial-discharge service can cause the batteries to develop a memory—that is, the batteries will only function properly up to the point where they are typically used. If batteries are used past that point, they might not produce the desired amount of energy for their circuit. Keep in mind that it takes a very long time for NiCad cells to develop a memory problem—it does not happen overnight.

A memory can sometimes be reversed by deeply discharging the battery, then recharging it through several complete discharge/recharge cycles. This might be accomplished by simply keeping the portable unit off the base during one or two days of normal use, then placing it on the base unit to recharge fully.

Finally, NiCad cells can simply wear out. Constant charging and discharging causes physical stresses in the battery itself and can eventually cause it to break down and become incapable of holding a substantial charge. When breakdown occurs, the battery pack should be replaced.

Range limitations

All cordless telephones are limited to some maximum working distance between the base and portable units. This distance limitation is based primarily on the level of transmit power used in your particular cordless telephone—greater power allows a greater operating range, and vice versa. Older models of cordless telephone could only work up to perhaps 50 yards or so. Circuitry was not efficient enough to develop more power without employing an unwieldy battery pack. Fortunately, the advances in battery design and electronic circuitry have extended portable operating range to 300 yards—even more in some of the newest units. It is doubtful that this range will be increased much more—range much beyond 300 yards can cause serious interference problems with other cordless telephones in adjacent homes or nearby apartments.

Radio transmitters are confined to a certain range because the strength of a signal decreases as an inverse square of the distance from the signal's source. This concept is known as the *inverse square law*. Although the mathematics and physics of this principle is beyond the general scope of this book, you should realize that signal strength does not vary in direct proportion to distance, but as:

$$[1/d^2]$$

where d is distance

For example, suppose you measure a radio signal 10 yards from its source, then measure the same signal 20 yards from its source. You might think that because the distance is twice as far, the signal should be half as strong—not true! The signal measured at twice the original distance would be only ¼ as strong. If you double the distance again to 40 yards and repeat your measurement, the signal will only be ¹⁄₁₆ as strong as it was originally. As you might imagine, it does not take long for a signal to fall off substantially.

Radio interference

Interference is any unwanted signal or condition that interacts with the desired signal and prevents the desired signal from reaching a destination. Interference can be caused by electrical means (i.e., a nearby cordless telephone or other piece of electronic equipment that generates RF energy), or by physical means (i.e., walls, doors, or partitions). While RF energy tends to penetrate nonmetal items with little problem, there is some amount of signal attenuation. Large metal objects and building structural metals have a much more pronounced effect on signal strength than nonmetal materials such as plastic or wood. Range is also related to interference—a weaker, distant signal is much easier to interfere with than a stronger, nearby signal

Very often, radio interference can be reduced or eliminated simply by moving the location of your cordless telephone—primarily the base unit. Try another room or part of the house. Try placing the base unit in a more open area away from televisions, computers, kitchen appliances, etc.

Radio privacy

Always keep in mind that your cordless telephone is a radio transmitter. The link between your base and portable units is made up of the public airwaves. Anyone with a receiver tuned to either your transmit or receive frequencies can hear at least half of your conversation (transmit and receive takes place on two different frequencies, so an eavesdropper cannot receive both sides of your conversation simultaneously). Still, potential eavesdropping is a great concern among privacy-minded individuals.

However, not even the most sophisticated receiver can pick up your signals beyond your cordless telephone's ability to transmit. An eavesdropper would have to be very close by to overhear you clearly.

To combat the possibility of radio eavesdropping, the new generation of cordless telephones employ digital signal processing to scramble the transmitted signal and descramble the received signal. Both the base and portable units contain compatible scrambler/descrambler circuits—and a security code pattern can be set to configure the units to work together. Any receiver not employing the necessary decoding pattern will receive only gibberish.

Troubleshooting cordless telephones

While much of the IC technology used in electronic telephones is also found in cordless telephones, cordless devices are more greatly complicated because of their use of precision radio circuits. To maintain clear, reliable radio communication, cordless telephones are set up and aligned precisely at the factory. Careless or indiscriminate adjustments can seriously impair the cordless telephone's radio performance. In general, transmitter and receiver alignment are not terribly difficult procedures, but you would require a broad array of test equipment such as an AF signal generator, an RF modulated signal generator, an RF frequency counter, an AF SSVM (single-sideband voltmeter), an oscilloscope, etc., to perform the alignment to factory specifications. The exact pieces of equipment required and alignment specifications will vary from cordless product to cordless product. With these difficulties in mind, this book does not cover cordless telephone alignment, or involved radio repairs. Instead, this chapter concentrates on simpler component check and replacement procedures.

Symptom 1   The ring signal in the base unit is low or absent. If your cordless telephone is not ringing at all, you must first make sure that a valid ringing signal is present on the telephone line to begin with. If another telephone on the same line rings correctly, it is a good indication that the trouble lies in your cordless telephone itself. If none of your available telephones will ring, the trouble might be in your home wiring, or in the local loop. You might want to troubleshoot your home wiring, as described in chapter 5.

When you are assured that a valid ringing signal is reaching your cordless telephone, check the telephone's modular connector very carefully to be sure that none of the connector's pins are bent or broken. Should you discover a damaged pin, replace the line cord connector.

Use your logic probe and measure the logic signal at the CPU's bell input during a regular ring cycle as shown in Fig. 9-5. You should find a logic 1 reading while the ring is in progress, and a logic 0 reading while the ring is idle. In some designs, this logic might be reversed, or you might read a clock signal. In any case, the CPU detects this change of logic state and transmits a control signal to the portable unit that causes it to ring.

If the ring logic signal at the CPU does not vary with the ringing pattern, it is possible that the ring detection optoisolator in the base unit might have failed. Use your multimeter to measure voltage across the optoisolator's input. This voltage should shift on and off in accordance with the normal ringing pattern. If this voltage varies, but the optoisolator output remains constant, replace the defective optoisolator.

If the optoisolator's input remains unchanged during a ring cycle, the circuit's connection to tip and ring is probably broken, or one of the dc blocking capacitors has become an open circuit. Remove all telephone power and check for any breaks in the printed circuit board or wiring. You can use your multimeter to check the integrity of each blocking capacitor, as described in chapter 4.

When the ring signal reaches the base unit's CPU as expected, but no ringing takes place, the CPU itself might be partially or totally disabled. Use your logic probe to measure the CPU clock signals generated at the oscillator crystal as shown in Fig. 9-6. If the

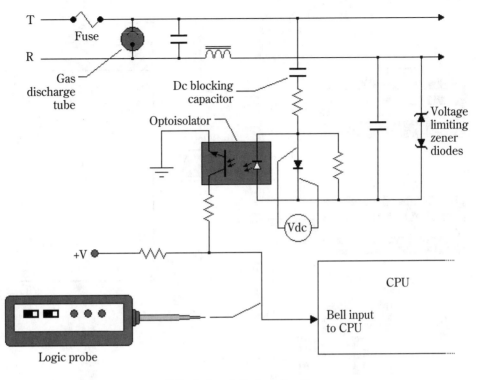

9-5 A ring detection circuit.

CPU's clock is running properly, your logic probe should indicate a clock or pulse response on both sides of the crystal. Make certain that your logic probe is capable of operating at the CPU's required clock speed.

If you detect only steady-state logic levels, the clock has probably failed. Remove all power from the circuit and try replacing the suspect crystal. Be sure to replace the crystal with only an exact replacement. If a new crystal does not restore operation, replace the CPU. If space permits, install an IC holder before inserting the new CPU.

Symptom 2   The base unit does not receive incoming calls. Begin by checking the hook switch relay as shown in Fig. 9-7. When the portable unit orders the base unit to pick up the telephone line, the base unit's CPU provides a logic signal to a transistor-driver circuit that operates a relay. In this application, the relay contacts act as a hook switch that completes the speech circuit and draws loop current from the central office.

Use your logic probe to test the logic signal operating your transistor driver. When the talk button is pressed on a portable unit, the relay driver signal from the CPU should become logic 1—or logic 0, depending on the particular circuit. If the logic signal does not shift, replace the defective CPU. If the driver logic appears to operate correctly, but the hook switch does not engage, replace the relay driver transistor(s) and the hook switch relay assembly.

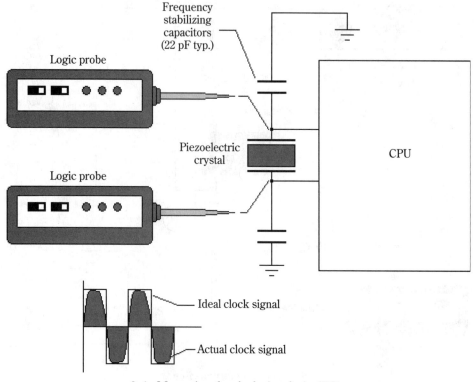

9-6   Measuring the clock signal of a CPU.

If the hook switch engages properly, remove all power from the telephone and use your multimeter to measure continuity across any small voice coil transformers. A shorted or open winding can easily disable the speech path into the telephone. If your particular telephone uses a speech network IC instead of voice transformers, try replacing the speech network IC.

If your have an oscilloscope with a transformer-coupled probe, you can trace the voice signal into the base unit right from tip and ring. Measure voice signals into and out of the base unit's AF amplifier stages. If voice should disappear at any one of these stages, replace the AF amplifier IC or transistor, then retest the cordless telephone.

RF carrier must be modulated with AF signals in order to achieve radio transmission. If the carrier frequency generated by the transmit (Tx) oscillator is not correct, the modulated AF signal will be filtered out by the portable unit's receiver instead of being passed and demodulated. This type of mismatch error can happen if the channel selections for the base and portable units are not the same. Check your channel selections and make sure that both units are set to operate with the same frequency pair.

Finally, check the *transmit mute* logic signal from the CPU. When the mute signal is true, the base unit will not transmit. A faulty logic signal from the CPU can inhibit the base unit's operation. If the CPU's transmit control signal is not correct, replace the CPU and retest the telephone.

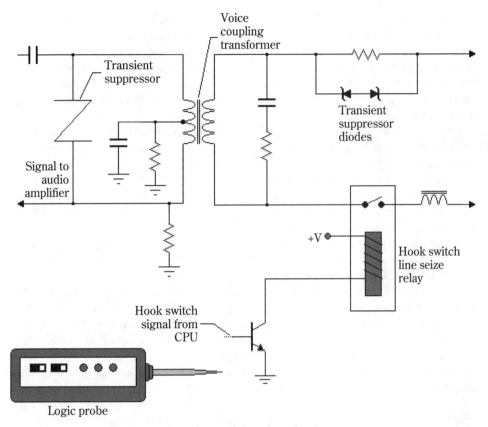

9-7 A speech interface circuit.

Symptom 3 The base unit cannot make or receive calls. In order for the base and portable units to operate as a set, each must have their transmit and receive frequency pairs set properly. This means both units must be set to the same channel. If the channel settings do not match, the base unit and portable unit cannot listen to one another. Speech and control signals will not be received by either unit. This is a surprisingly common problem (usually due to user error) and can easily disable the entire telephone. Begin your investigation by checking the channel selections for the base and portable units. Make sure that each is set for the same channel.

If the base unit is active (that is, it receives power and responds to an incoming ring), but will not seize the telephone line, you should inspect the hook switch relay as in Fig. 9-7. Faulty contacts in the relay or a missing logic signal operating the relay's transistor driver can prevent the base unit from accessing the telephone line. Use your logic probe to measure the CPU output that operates the relay's transistor driver. If the logic signal does not shift when base unit should go offhook, the CPU might be faulty. Try replacing the CPU and retest the circuit. When the CPU's relay signal shifts as expected, but the relay does not engage, replace the hook switch relay and its driver transistor(s).

Beyond a mistaken channel setting or malfunctioning hook switch relay, it could be that the communications path between your base and portable units is not working. The proper method of testing would be to trace the speech signals in both units. If either unit does not transmit or receive at any point in the circuitry, that is the probable location of the failure. Keep in mind that you will probably need an assortment of RF test equipment and a wide-bandwidth oscilloscope to trace the speech signals in your cordless telephone.

Symptom 4   The base unit cannot make outgoing calls for this procedure; refer to the block diagram of Fig. 9-3. In many cases, a failure has occurred in the base unit's receiver circuit that prevents speech and control signals from reaching the telephone line. Use your oscilloscope to measure AF signals in the receiver circuit. If no AF signals are being received, the receiver circuit is defective. You will need a selection of RF test equipment to troubleshoot the receiver. The point where the received signal disappears is the most likely point of failure.

Use your logic probe to inspect the receiver mute line from the CPU. If the mute signal is incorrect (for example, it keeps the base unit's receiver idle all the time), the CPU might be defective. Try replacing the CPU and retest the base unit.

Symptom 5   Audio signals in the portable unit are always noisy. For this procedure, refer to the block diagram of Fig. 9-4. Faulty or loose wiring, a hairline fracture in the PC board, or a defective local oscillator or signal detector can easily cause troublesome noise. Before you disassemble your portable unit, make sure that both units are near each other and operating on the same channel. Try the cordless telephone in a new location or set both units to a different operating channel—outside interference can sometimes cause noisy operation.

Carefully inspect any jumpers and interconnecting wires that connect the antenna, speaker, or any parts of the portable unit's PC board(s). Secure or resolder any wires that appear loose or frayed. If your noise problem began abruptly after the portable unit had been dropped or banged, you could have a hairline fracture somewhere in the PC board. If you tap on the portable unit, you might hear extra static—a clear indication of PC board trouble. Refer to chapter 3 for service guidelines on PC board repair.

If noise persists, use an RF frequency counter to measure the portable unit's local oscillator frequency. If the local oscillator's frequency drifts, it might reduce the signal detector's ability to discriminate the desired audio signal. This could introduce severe distortion into the received signal. If you are able to measure the local oscillator frequency and know what it should be according to manufacturer's specifications, you can adjust the local oscillator's frequency.

If you absolutely must adjust the local oscillator without instruments or instructions, a procedure that is strongly discouraged, place a mark on the adjustment with an indelible marker to indicate your starting place, then carefully count the number of turns clockwise or counterclockwise. Advance slowly, using ¼-turn increments (or smaller). If you get into trouble, at least you will have a marked place to go back to. If the local oscillator has drifted, it should take less than one turn either way to correct the frequency. If you must turn the oscillator adjustment more than one turn, your trouble probably lies elsewhere.

Symptom 6   The portable unit does not go offhook when the talk button is pressed. When the talk button is pressed, a control code is transmitted to the base unit which, in turn, causes the base unit to seize the available telephone line. The talk signal is interpreted by the portable unit's keypad decoder, and an RF output is enabled. If you have an RF frequency counter, press the talk key and be sure that RF is being provided to the portable unit's antenna. If no RF is present, the trouble is probably in the portable unit's transmitter circuit or CPU circuit. Use your logic probe to check the transmit mute logic signal from the portable unit's CPU. If the transmit mute signal is incorrect, replace the portable unit's CPU.

If the RF signal is available as expected, it might be that no control codes are being generated from the portable unit. Replace the AF amplifiers in the transmit circuit and retest the unit. If these steps fail, the trouble could lie in your base unit, so refer to symptoms 3 and 4.

Symptom 7   The portable unit does not ring. This symptom is a continuation of Symptom 1. If you have already inspected the base unit and found it to be operating properly, you should suspect a fault in the portable unit. The easiest way to begin is by checking the portable unit's ringer transducer (speaker or a piezoelectric element). Whatever type of signalling device is at work, check all connectors and interconnecting wiring to be sure that the signalling device is connected properly. Replace any damaged signalling device.

When the base unit rings, a control code is transmitted to the portable unit. Use your oscilloscope or logic probe to measure the ringer control signal reaching the portable unit's CPU. If this signal is missing, the control code detection circuit in the portable unit might be defective. Measure the ringer signal being generated by the portable unit's CPU. If the CPU is not producing a signal to activate the ringer, then the CPU might be defective. Try replacing the CPU. If the CPU does produce a ringing signal, but the ringer does not ring, the circuit that generates the audible ringing signal might be defective.

Symptom 8   Speech is not transmitted from the portable unit. This type of symptom prevents you from making outgoing calls. You might be able to hear normal operation on your receiver, but you cannot dial or speak. Check your portable unit's microphone and microphone wiring to ensure that nothing is damaged or loose.

Your next step is to check the transmitter mute signal being generated by the CPU. When the portable unit is offhook (the talk button is pressed), you should read a logic 1 on the transmit mute pin to indicate that the portable unit's CPU has released the transmit circuit. This should make speech and DTMF signals available to the portable unit's transmitter circuit. If you measure a logic 0 at the transmitter mute pin, the CPU is disabling the portable unit's receiver. The CPU might be defective. Try replacing the portable unit's CPU and retest the telephone.

Finally, you can use your oscilloscope to trace your transmitted speech through the portable unit's AF amplifier circuits. Trace the AF signals into the transmitter circuit. If speech signals disappear before reaching the transmitter, the point at which signals disappear is probably the point of failure. If speech signals can be traced into the transmitter circuit, the transmitter circuit itself might be defective.

Symptom 9   Speech cannot be heard on the portable unit. Under most circumstances, a loss of signal reception at the portable unit is complete—you will not hear dial tone or speech received from the base unit. Begin your repair by inspecting the portable unit's receiving speaker. Make sure that the speaker is intact, along with its wiring and connectors. Secure any loose or damaged wiring. Replace the speaker if it appears damaged.

Your next step is to check the receiver mute signal being generated by the portable unit's CPU. When the portable unit is offhook (the talk button is pressed), you should measure a logic 1 on the receiver mute pin. A logic 1 indicates that the CPU has released the receiver circuit, and speech signals should be made available to the portable unit's AF amplifier circuits. If you measure a logic 0 at the receiver mute pin, the CPU is disabling the portable unit's receiver. The portable unit's CPU might be defective. Try replacing the portable unit's CPU and retest the telephone.

Finally, you can use your oscilloscope to trace AF signals available from the portable unit's receiver circuit. You should be able to trace received AF signals all the way to the speaker in the portable unit's earpiece. If speech is available from the receiver, but does not reach the earpiece, the point at which the AF signal disappears is the probable point of failure. Replace any components that you believe to be faulty.

Symptom 10 The portable unit does not charge. The base unit provides a charging voltage for its portable unit through a set of open contacts in the base unit. When the portable unit is idle and placed into the base unit's cradle, the base unit's contacts mate with the corresponding contacts on the portable unit. This system effectively connects the portable unit to the base unit. An LED indicator is often provided on the base unit to show when charging current is flowing to the portable unit.

If the charging indicator does not activate when the portable unit is resting in its cradle, clean and check both sets of contacts to be sure that they are making proper contact. Reseat the portable unit to be sure that it is resting properly. If the charging lamp still does not illuminate, use your multimeter to check the dc charging voltage across the base unit's contacts.

Keep in mind that the base unit's voltage output might be interlocked to a weight-sensitive switch in the cradle—voltage will not be made available unless there is something in the cradle to turn on the switch. You can press down on such a switch by hand, if necessary, while making your measurements. If charging voltage is low or absent, your trouble probably lies in the base unit's charging circuit. A regulator in the charging circuit might be defective. Use your multimeter to trace dc charging voltage through the charging circuit.

When the charging indicator does activate, your trouble probably lies in the portable unit's battery pack. Nickel-cadmium batteries can fail after long periods of continuous use. In most designs, the charging voltage from the base unit is routed directly across the battery pack, so there is rarely any charging circuitry within the portable unit. Disconnect the battery pack and measure the charging voltage at the open battery terminals. If this voltage is low or absent, your charging filter capacitor might be shorted, or the charging rectifier diode may be open. Remove power and check these components as detailed in chapter 4. Replace any components that appear to be defective.

Symptom 11 The telephone's operating range is too short. As you might have seen earlier in this chapter, a cordless telephone's operating range is a function of its transmitted power, and any sources of interference that can exist between the base and portable units. Begin your investigation using your multimeter to measure the dc voltage being supplied at the base unit's power input. Most models of cordless telephone use a small, self-contained power pack to provide the base unit with power, so it is a simple matter to unplug the dc power connector and measure dc input power directly. Your voltage readings should closely match the rated output marked on the power pack. If dc input power is low, the base unit will not receive enough energy to transmit correctly, or supply ample charging power to the portable unit.

Consider the portable unit itself. If the portable unit has been in service for a while without an ample charge, its battery pack might be too depleted to transmit signals back to the base unit. Make sure the portable unit is properly charged.

Beyond the considerations of power, you should be primarily concerned with sources of interference. Nearby RF-generating equipment, such as that used by amateur radio operators and citizens band enthusiasts, can be prime sources of interference. High-energy equipment, such as heavy-duty motors, microwaves, arc lamps, welding equipment, etc. can also provide substantial interference. Should you find such interference sources, try moving your base unit to other locations in your home or office—or, if possible, remove the source(s) of interference. You could also try selecting a different channel for the base and portable units. A new channel might be more immune from interference sources.

10

Cellular telephones

THE ABILITY TO TALK WHILE ON-THE-GO HAS ALWAYS BEEN A SOUGHT-AFTER luxury. For many users, mobile communication offered both convenience and efficiency—people could be reached anywhere at any time. For other users, mobile communication proved attractive as a means of emergency communication. These considerations are even more prevalent today.

Mobile communication has been available for almost sixty years in the form of radio technology. Amateur radio and citizen's band technologies enjoy widespread acceptance and participation. However, there is virtually no access to the PSTN using broadcast radio equipment. Some amateur radio equipment does provide a link to the PSTN, but a lengthy procedure is needed to use the link, and it can only be used by a single operator at any one time—and then, only for brief periods.

Eventually, a line of radiotelephone equipment was designed that allowed direct radio links with the PSTN from automobile-based radio sets. In principle, the radio-telephone operates much like today's cordless telephones as covered in chapter 9. The portable unit would be in your car, and a large number of base units would be located at regional radio facilities. Such an arrangement allowed a number of users to make calls simultaneously—each user communicated on a different set of frequencies (or channel). The physical equipment needed to interface to the PSTN was located at a central radio facility.

Radiotelephone technology was limited by two very important factors. First, there are only just so many channels available in the RF band (range of frequencies) allocated by the FCC for radiotelephone use. The standard radiotelephone band supported up to 2000 channels. While this might sound like a large number of channels, there simply were not enough to support the large number of people who desired service—

waiting lists for radiotelephone service were outrageous. Second, radiotelephone was a centralized service—you subscribed to a local service, and you received service from that provider only. Range was limited by the provider's radio facilities.

Cellular telephone service was developed to offer convenient telephone communication over a vast area at a very affordable price to everyday users. This chapter examines cellular telephone technology, and explores the inner workings of a typical cellular telephone set (Fig. 10-1).

10-1 A pocket-sized cellular telephone.

The cellular approach

The biggest problem with conventional radiotelephony was that it used a brute force approach to provide area coverage. Regional stations provided as many channels as possible with as much power as the Federal Communications Commission (FCC) allowed. This approach quickly resulted in very limited islands of service. Each island was separated from another by substantial geographic distances to prevent mutual interference between similar channels in use. Increased station power and wider frequency band allocations from the FCC were not considered feasible alternatives for improving the performance of radiotelephony. Instead, the entire mobile telephone approach had to be rethought from the drawing board.

The American Telegraph and Telephone Company (AT&T) was the first to address the redesign of mobile telephony by developing the Advanced Mobile Phone Service (or AMPS)—known today simply as the *cellular system*.

Cellular structure

The cellular system represents an innovative and radical departure from conventional radio communication as shown in the illustration of Fig. 10-2. Instead of establishing a single, high-power communication path between two points, the cellular system divides a geographical region into relatively small areas (or *cells*). A station in each cell is equipped with a low-power radio system, along with computer-controlled equipment that links each cell to a centralized mobile telephone switching office (MTSO). End users do not communicate directly with an MTSO—they communicate instead with their nearest cell station. Each cell site then communicates with a region's MTSO. In this way, a network of many interdependent radio stations can be created over an extremely broad area.

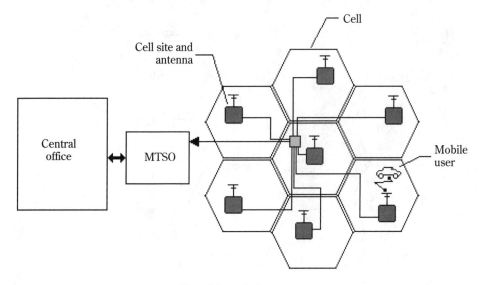

10-2 The cellular concept.

The revolutionary aspect of cellular communication is that each cell is kept small—the area of coverage is only a few miles in diameter. With limited range, the same sets of frequency channels can easily be reused in nonadjacent cells. A single cellular telephone operating band can therefore provide virtually worldwide coverage. The use of low power also means that the telephone equipment can be kept extremely small and efficient.

In the United States, the FCC allocated a frequency band for cellular operation that allows 666 channels. A typical cell site is designed to handle up to 45 full-duplex conversations simultaneously. Each conversation requires two channels (frequencies) to achieve full-duplex operation, and a cell can use up to 90 of those 666 channels. Adjacent cells use other sets of channels, while nonadjacent channels can reuse the same sets of channels.

Suppose, for the example of Fig. 10-3, that the center cell of Area 1 uses channels 1 to 90. No cells adjacent to Area 1 can reuse those same channels because of the possible interference that would result, so the adjacent cells will utilize other channels within the 666 channel band. The cells of Area 2 are far enough away to reuse the same channels as Area 1. The MTSO accepts speech from each of its cell sites, as well as other regional MTSOs, and provides the actual connections to the PSTN.

When a cellular user requests service by taking their cellular telephone offhook,

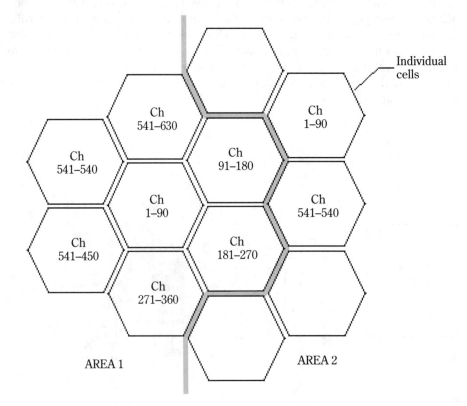

10-3   Typical channel distribution among cells.

the nearest cell site receives the telephone's transmission and locates two open channels through which to establish a link. The choice of channel is fully automatic—the user need only request service. Once a full-duplex link is established, the managing MTSO opens a telephone line to its local central office. You will hear dial tone when an open line is ready.

As the mobile telephone moves across a region, the signal strength between it and other adjacent cell sites is constantly changing. When a user gets too far away from one cell, the MTSO determines what cell is best able to take over the call. Control of the call is then transferred to the desired cell site in a process known as *handoff*. The entire handoff process is totally transparent to the user, and the ongoing call continues uninterrupted.

Another advantage to the cellular system is that various regions served by different cellular networks can be made to work together to establish a universal network. Many cellular service providers have entered into reciprocal service agreements with other cellular service providers. Whenever you travel outside of your regular service area, a competitor's network automatically accepts the handoff and provides you with service. Where two different service areas are adjacent to one another, service can continue uninterrupted between regions.

Finally, the cellular network is expandable. Not only can new cells be added to the existing network, but existing cells can be split and divided into an array of smaller cells to service larger numbers of users.

The cellular telephone

There is little doubt that cellular telephones are some of the most sophisticated and powerful communications devices currently available for consumers. For the purposes of this book, a cellular telephone can be broken down into three distinct modules as shown in Fig. 10-4: the radio (RF) module, the audio (AF) module, and the control/logic (CPU) module.

The RF module

The RF module handles all signals entering or leaving the cellular telephone, as shown in Fig. 10-5. An antenna, connected to a solid-state duplexer, acts to isolate transmitted signals so they cannot feed back into the receiver circuit. The duplexer's filtering action allows full-duplex communication—while operating the cellular telephone at up to several watts of transmitting power.

Received signals are filtered and demodulated by the RF receiver circuit. The RF module's output is supplied to the AF module. However, where conventional radio receivers use manual tuning to define a desired channel, the cellular telephone uses a precision frequency synthesizer circuit that can be set to any of the 666 allocated cellular channels. The channel selected at any point in time is determined by the control/logic module. As your cellular telephone moves from cell to cell, transmit and receive frequencies are switched to accommodate the available channels of the new cell. Instructions dictating which frequencies to switch to are received as data signals that are processed by a modem in the cellular telephone's control/logic module.

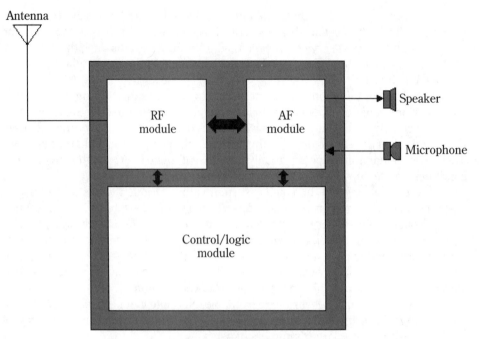

10-4   A cellular telephone.

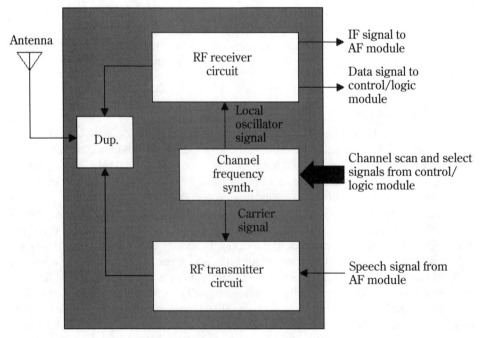

10-5   A cellular telephone's RF module.

Speech signals from the AF module and data signals from the control/logic unit are sent to the RF transmitter circuit—where signals are placed on the appropriate RF carrier, filtered, amplified, and supplied to the antenna. The RF carrier frequency is determined by the particular cell in which you are operating.

The channel-frequency synthesizer circuit generally consists of a base oscillator in conjunction with a receive-frequency synthesizer and a transmit-frequency oscillator. The receive-frequency synthesizer accepts a digital control signal from the control/logic module, then produces a voltage that is proportional to the desired frequency. A voltage-controlled oscillator, or VCO, converts the proportional voltage into an oscillator signal. A similar circuit is available for the transmitter-carrier circuit. Digital control signals from the control/logic module set a voltage that is proportional to the desired frequency. The proportional frequency drives a VCO—and produces the oscillator frequency.

The AF module

The AF module is responsible for converting IF (intermediate frequency) signals from the RF module into speech signals that can be heard in the cellular telephone's receiver, as shown in Fig. 10-6. A second receiver element is often included to produce any alarm signals, such as ringing signals. DTMF dialing tones and speech from a microphone are filtered, mixed together, then supplied to the RF module for modulation—along with control signals from a modem in the control/logic module. A portion of transmitted speech is returned to the receiver as sidetone. AF transmit and receive operations are under the direct control of the control/logic module.

The control/logic module

As you can see in the block diagram of Fig. 10-7, the control/logic module is the heart of a cellular telephone. A control/logic module resembles the architecture of a per-

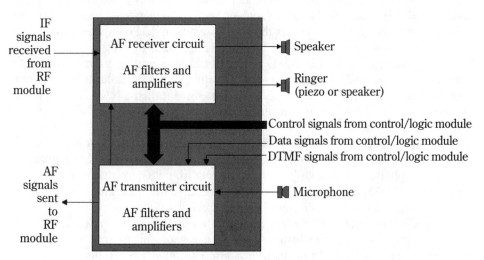

10-6 A cellular telephone's AF module.

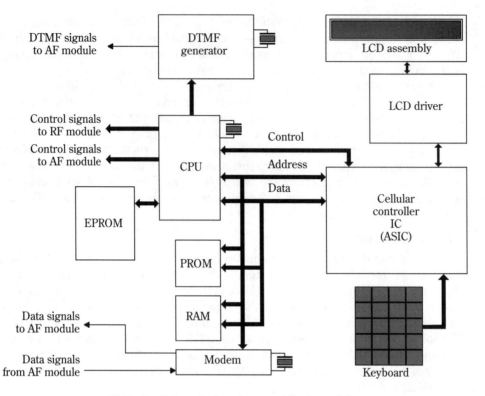

10-7   A cellular telephone's control/logic module.

sonal computer. The main CPU runs the cellular telephone based on a set of permanent instructions (its program) retained in permanent memory (ROM). A temporary memory (RAM) is included to hold variables such as the current channel, the transmitter power setting, etc. as well as the results of any logical comparisons or mathematical operations required while the telephone program is running.

An erasable memory (EPROM) is used to hold operating information that is unique to each individual telephone—such as the assigned cellular telephone number. This type of individualized memory is sometimes called a NAM, or number assignment module. The CPU is in direct control of the AF and RF modules, as well as the DTMF generator.

Because a cellular telephone is an active part of the cellular network, it must be in constant contact with the network. In addition to speech and DTMF signals, the cellular telephone must transmit and receive data from its current cell site (and ultimately, the MTSO). A modem IC is used to add data to the transmitted signal, and interpret commands and data delivered from the cellular network.

The CPU also manages the operations of a *cellular controller* IC. The cellular controller is typically a sophisticated ASIC that is responsible for interfacing with the cellular telephone's keypad and display system. The cellular controller sets the transmit and receive frequency synthesizers in the RF module.

A display is usually included in the cellular telephone to show the telephone number being dialed, along with the status of the network (i.e., Select, On, In-Use, Roam, No Service, etc.). LCD driver/display modules are the display of choice due to their very low power consumption and long working life. Numbers and messages up to 16 characters long can be displayed in the LCD.

Cellular troubles

As with cordless telephones, cellular telephones have several disadvantages that you should be aware of. Remember that disadvantages are not necessarily defects or flaws in a cellular telephone's design, but are only a part of the product's nature. In most cases, these disadvantages arise in the radio link between the cellular telephone and a cell site. Cellular troubles can be grouped into four general categories: dropouts, dead zones, battery problems, and radio privacy.

Dropouts

An inherent problem with radio signals in the 800- to 900-MHz range (the cellular communications band) is that signals tend to move only in straight lines away from your antenna. Such high-frequency radio waves are weakened, or attenuated, by moisture in the atmosphere, reflected by buildings and smooth surfaces such as water, and can be blocked entirely by large geographical obstructions like hills or mountains.

When your cellular telephone is in motion, received signal strength can fall off enough in some instances to cause brief interruptions in the received signal. More severe cases can interrupt your transmitted signal back at the cell site. You will notice such dropouts as sudden pauses in reception. There might be only one or two brief pauses, or a series of variable-length pauses, depending on the severity of the circumstances.

Another common cause of dropouts occurs when you are approaching the fringe region of a service area where there are no other cell sites to accept the handoff of your conversation. You will experience a gradual weakening of the signal until brief dropouts begin. Dropouts will quickly worsen until you are totally disconnected.

Cell site controls are generally designed to disregard minor dropouts without interrupting your conversation. Continuous or prolonged dropouts, however, can cause the cell site to disconnect you. Experience will teach you where the poor areas of coverage are in your region.

Dead zones

In principle, dead zones occur for the same general reasons as dropouts—but the area of poor coverage is on a much larger scale. Received signals can drop out for so long that the cell site interprets the signal loss as a hang-up. A cell site responds by clearing the lost channels, and reassigning the channels as needed by other calls.

Hilly, mountainous, or dense urban areas often experience dead zones. Signals are absorbed or reflected, preventing radio waves from traveling to the desired area. A dead zone can sometimes be eliminated by changing the location of a cell site, or by splitting the cell to add extra sites that will properly cover the afflicted area.

Battery problems

Cellular telephones are powered by rechargeable battery packs made up of several NiCad (nickel-cadmium) batteries. While NiCad batteries offer an effective and convenient method of powering the telephone, the batteries do suffer from several drawbacks that you should be aware of.

First, NiCad batteries offer a somewhat lower energy density than non-rechargeable batteries. Because their energy density is relatively low, NiCad batteries are not well suited for providing energy to heavy loads, or loads exerted for prolonged periods of time (without recharging). In fact, NiCad cells will eventually go dead just sitting on a shelf unless they receive a constant standby, or trickle charge. Although the materials and construction of NiCad batteries has improved, and sophisticated ICs have lessened the overall power demands on NiCad batteries, you should not expect more than a few hours of service from a set of NiCad batteries before they require recharging. Fortunately, few calls take that long, so it is usually convenient to keep the cellular telephone in a charging station when not in use.

NiCad batteries can also become troublesome when they are regularly discharged to the same levels and then recharged again. This can happen, for example, if you spend an average of 30 minutes per call several times a day, and you allow the telephone to recharge between calls. Such partial-discharge service can cause the batteries to develop a *memory*—that is, the batteries will tend to function properly only up to the point where they are normally discharged. If batteries are used past that point, they will not have the required (or expected) amount of energy to operate the circuit.

Keep in mind that it takes a very long time for NiCad batteries to develop this type of problem—it does not happen overnight. A memory can sometimes be reversed by deeply discharging the battery, then recharging it through several complete cycles. This might be accomplished by simply keeping the telephone off its charging station for one or two days of normal use, then letting the telephone recharge fully.

Finally, NiCad batteries simply wear out. Constant charging and discharging cause physical stresses in a battery that can eventually cause it to break down and become incapable of holding a substantial charge. When this occurs, your best course is to replace the battery pack entirely.

Radio privacy

It is important for you to realize that your cellular telephone is largely a radio transceiver. The link between your cellular telephone and the nearest cell site is made up of the public airwaves. As a consequence, anyone with a receiver tuned to either your transmit or receive frequency channel will be able to hear at least half of your conversation. Transmit and receive operations take place on two different frequencies, so an eavesdropper cannot listen to both sides of a conversation simultaneously.

This has become a cause of great concern among privacy-minded individuals. However, not even the most sophisticated receiver can receive signals beyond your ability to transmit. Cellular telephones typically have a range of several miles, so an eavesdropper would have to be nearby in order to receive you clearly. Also, when a cellular telephone is in motion, the conversation switches channels as handoff takes place between cells. An eavesdropper would have to follow you, and be able to search each of 666 channels for your conversation—a virtually impossible procedure for even the most skilled radio professionals.

To combat the remote possibility of electronic eavesdropping, a new generation of cellular telephone accessories uses digital signal processing and compression techniques to scramble your transmitted voice—and descramble your received voice at the destination telephone. A caller at the other end of your conversation must also have a similar accessory set to the same security pattern. Any speech signals transmitted over the public airwaves would be scrambled—unintelligible to anyone that might be listening without a properly configured descrambler.

Troubleshooting cellular telephones

In spite of their small size, cellular telephones are remarkably sophisticated devices—part telephone, part radio, and part computer—all combined into a device that fits into the palm of your hand. As a result, it is virtually impossible to present a useful troubleshooting guide for cellular telephones in a single chapter. Too much of cellular troubleshooting relies on a knowledge of RF and computer technologies that are simply beyond the scope of this book.

If you choose to attempt the repair of a cellular telephone, it is vital that you have a service manual for your particular model, and that you have the necessary test equipment to repair and align the telephone properly. A service manual lists the manufacturer's part numbers for ICs (many ICs are proprietary to the manufacturer) and other components that you might need to order as replacement parts. A service manual will also provide you with a complete set of schematic diagrams which are vitally important when dealing with a device as complex as a cellular telephone.

Appendix A
Troubleshooting charts

Chapter 6
Classical telephones

The telephone is completely dead. No side tone or CO dial tone is audible in the receiver.

√ Confirm that the telephone is defective.
√ Check the voltage across the local loop.
√ Check/replace the handset assembly.
√ Check/replace the speech network assembly.
√ Check/replace the hook switch assembly.

Ring is low (weak) or absent.

√ Check the ringer volume control.
√ Check/replace the ringer assembly.
√ Check/replace the dc blocking capacitor.

Noise is present in the transmit or receive audio.

√ Confirm that the telephone is defective.
√ Check/replace the handset cord and its connections.
√ Check/repair connections on the speech network PB board.
√ Check/clean the hook switch contacts.

Rotary dial problems. Rotary dial will not break dial tone once dialing pulses are sent.

√ Clean rotary dial contacts.
√ Check/replace the rotary dial assembly.

DTMF keypad is unresponsive. No tones (or only one tone) generated when a key is pressed, or the key must be depressed firmly to engage both tones. (Classical DTMF dial units only.)

√ Clean DTMF dial contacts.
√ Check/replace the DTMF dial assembly.

Low or distorted transmission from the telephone.

√ Check/replace the carbon microphone element.

Low or distorted reception from the telephone.

√ Check/replace the electromagnetic receiver element.

Chapter 7
Electronic telephones

The telephone is completely dead. No sidetone or central office dial tone is audible in the receiver (or speakerphone output).

√ Check/replace the telephone's power supply (ac/dc adaptor).
√ Check/replace the telephone's line cord.
√ Check output from the polarity protection bridge.
√ Check/repair any defective wiring or connectors.

Speakerphone functions do not work in transmit and/or receive modes.

√ Check/repair connections at the speaker or microphone.
√ Replace the speakerphone circuit IC.
√ Check/replace any defective transistors in the speakerphone speech network.
√ Check/replace any defective voice transformers in the speakerphone speech network.

The handset does not work in transmit or receive modes.

√ Check/replace the handset cable.
√ Check/repair connections at the handset.
√ Check/replace the defective microphone or receiver.
√ Check/replace the defective hook switch.

The visual display is absent or erratic.

√ Check/repair faulty wiring or connections at the display.
√ Check/replace defective CPU crystal(s).
√ Replace the defective CPU.
√ Replace the defective display-driver IC.
√ Replace the defective display.

Pulse dialing does not work on one or more keys.

√ Check the position of the pulse/tone switch.
√ Check/repair defective wiring or connectors within the telephone.
√ Check/clean dial pad keys.
√ Check/replace defective transistor(s) in the loop pulsing circuit.
√ Replace the defective CPU.
√ Replace the defective dialer IC.

Tone (DTMF) dialing does not work on one or more keys.

√ Check the position of the pulse/tone switch.
√ Check/repair defective wiring or connectors within the telephone.
√ Check/clean dial pad keys.
√ Check/replace the defective clock crystal at the CPU or dialer IC.
√ Replace the defective CPU.
√ Replace the defective dialer IC.
√ Check/replace any switching transistor(s) in the DTMF circuit.

Ringing is faint or absent.

√ Check the ringer volume control.
√ Check/replace the defective dc blocking capacitor.
√ Check/replace ring coupling transformer (if any).
√ Replace the defective ringer IC.

No redial or number storage functions.

√ Check/clean keyboard contacts.
√ Replace the defective CPU.
√ Replace the defective memory IC.

Chapter 8
Answering machines

Message tape does not move at all (ICM or OGM).

√ Check/replace the defective tape(s).
√ Check/reseat the flywheel belt.
√ Check/repair the defective tape mechanism.
√ Check/replace any defective motor-driver transistor(s).
√ Replace the defective motor-driver IC.

No audio is processed during tape play (ICM or OGM).

√ Check/replace defective play/record amplifier IC.
√ Check/replace defective magnetic play/record head.
√ Check/replace defective speech network IC.
√ Check the speaker and associated wiring.
√ Replace the defective speaker amplifier IC.

Machine does not pick up on incoming ring.

√ Check/replace defective dc blocking capacitor.
√ Check/replace defective zener protection diodes.
√ Check/replace ring detection optoisolator.
√ Check the speaker and associated wiring.

No beep tone is generated (or played back) at the end of messages.

√ Replace the defective CPU.
√ Replace the defective magnetic play/record amplifier IC.

A tape does not rewind.

√ Check/repair the defective tape drive mechanics.
√ Check/replace the defective motor-driver transistor(s).
√ Replace the defective CPU.

OGM is heard through the speaker, but not transmitted to the telephone line.

√ Check/replace the defective speech network IC.
√ Check for circuit interruptions to the telephone line.

New ICMs are mixed together with old ICMs (old ICMs are not being erased).

√ Replace the defective magnetic erase head.
√ Replace the defective erase head driver IC.

Machine disconnects the telephone line in only a few seconds after an ICM starts.

√ Check/replace the defective vox IC.

A new OGM will not record (tape does move).

√ Check/replace the defective microphone or associated wiring.
√ Check/replace the defective magnetic play/record head.
√ Check/replace the defective play/record amplifier IC.

No control is provided by remote.

√ Tone decoder(s) are maladjusted.
√ Replace defective tone decoder(s) in the remote circuit.
√ Replace the defective DTMF amplifier.

The machine does not release the telephone line when a caller hangs up.

√ Confirm that the CO is generating a CPC pulse.
√ Check the setting (long or short) of the CPC switch.
√ Check/replace the defective CPC diode bridge.
√ Check/replace the defective CPC optoisolator.
√ Check/replace the defective optoisolator driver transistor(s).

Chapter 9
Cordless telephones

The ring signal in the base unit is low or absent.

√ Check the telephone line at the base unit.
√ Check/replace the ring detection optoisolator.
√ Check/replace the CPU clock crystal.
√ Replace the defective CPU.

The base unit does not receive incoming calls.

√ Check/replace the defective hook switch relay.
√ Check/replace defective audio coupling transformer(s).
√ Check the audio signal at the main audio amplifier.
√ Check the base and portable unit channel settings.
√ Check the transmit output from the CPU.

The base unit cannot make or receive calls.

√ Check the base and portable unit channel settings.
√ Check/replace the defective hook switch relay.
√ Trace the speech signals through the base unit.

The base unit cannot make outgoing calls.

√ Check for audio output from the IF amplifier.
√ Check for audio output from the AF amplifier.

√ Replace the defective CPU.
√ Check/replace the defective audio coupling transformer(s).

Audio signals in the portable unit are always noisy.

√ Select a new channel for the base and portable units.
√ Test the telephone in a new location.
√ Check/repair any faulty connections on the PC board.
√ Check/adjust the local oscillator signal.

The portable unit does not go offhook when the talk button is pressed.

√ Check for the presence of an RF signal at the portable unit's antenna.
√ Check/replace the defective CPU.
√ Check/replace the defective AF amplifier IC(s).
√ Confirm the operation of the base unit.

The portable unit does not ring.

√ Check the alarm speaker and its associated wiring.
√ Check/replace defective code-detector IC(s).
√ Check/replace the defective CPU.
√ Check/replace the defective ring oscillator.

Speech is not transmitted from the portable unit.

√ Check/replace the microphone.
√ Check/replace the defective CPU.
√ Trace audio signals through the transmit circuit.

Speech cannot be heard on the portable unit.

√ Check/replace the audio speaker.
√ Check/replace the defective CPU.
√ Trace audio signals through the receive circuit.

The portable unit does not charge.

√ Check/clean charging contacts at the base and portable units.
√ Check the charging circuit in the base unit.
√ Replace the portable unit's battery pack.

The telephone's operating range is too short.

√ Check the charging voltage at the base unit.
√ Select a new channel for the base and portable units.
√ Test the telephone in a new location.

Appendix B
Parts and supplies vendors

Active Electronics
Various locations across the United States
(800) 677-8899

Active stocks many commercial passive and active electronic components, along with books, tools, and supplies. Call for their catalog and nearest location. Major credit cards accepted.

Consolidated Electronics, Inc.
705 Watervliet Avenue
Dayton, OH 45420
(800) 325-2264
(513) 252-5662

Consolidated stocks a wide assortment of electromechanical and electronic parts by domestic and foreign manufacturers, as well as test equipment, tools, and supplies. Most major credit cards accepted.

Dalbani Corporation
2733 Carrier Avenue
Los Angeles, CA 90040
(800) 325-2264
(213) 727-0054

Dalbani stocks a wide variety of electromechanical and electronic parts by domestic and foreign manufacturers, as well as mechanical parts. Call for their catalog.

Stephen J. Bigelow
Dynamic Learning Systems
P.O. Box 805
Marlboro, MA 01752

The author will be happy to answer your technical questions, provide additional information, and receive your comments and experiences. Feel free to write anytime.

Howard W. Sams & Company
2547 Waterfront Parkway East Drive
Indianapolis, IN 46214
(800) 428-7267

Sams offers an impressive selection of service documentation for a large number of consumer electronics products. Call for their complete documentation catalog.

MCM Electronics
650 Congress Park Drive
Centerville, OH 45459-4072
(800) 543-4330

MCM supplies a large number of foreign ICs and semiconductors, as well as electromechanical parts, tools and supplies. Call for their catalog.

NTE Electronics, Inc.
44 Farrand Street
Bloomfield, NJ 07003
(800) 631-1250

NTE provides a comprehensive selection of aftermarket active and passive electronic devices—including many foreign components. Call for their complete catalog.

Glossary

ADC (Analog-to-Digital Converter) An electronic device used to convert an analog voltage into a corresponding digital representation.

AF (Audio Frequencies) The frequencies that fall within the range of human hearing, typically 50 to 18,000 Hz.

AM (Amplitude Modulation) A technique of modulating a carrier sinusoid with information for transmission.

anode The positive electrode of a two-terminal electronic device.

attenuation The loss or reduction in a signal's strength due to intentional or unintentional conditions.

bandwidth The range of frequencies over which a circuit or system is capable of operating, or is allowed to operate.

base One of three electrodes of a bipolar transistor.

battery The operating voltage supplied to a telephone from a central office.

BOC (Bell Operating Company) The local telephone company that provides your telephone service from your central office.

capacitance The measure of a device's ability to store an electric charge, measured in farads, microfarads, and picofarads.

capacitor A device used to store an electric charge.

cathode The negative electrode of a two-terminal electronic device.

cell In cellular telephony, the geographic area served by one transmitter/receiver station.

channel An electronic communication path. A channel may consist of fixed wiring or a radio link. A channel has some bandwidth, depending on the type and purpose of the channel.

CO (Central Office) The building and electronic equipment owned and operated by your local telephone company that provides service to your telephone.

collector One of three electrodes on a bipolar transistor.

continuity The integrity of a connection measured as very low (ideally zero) resistance by an ohmmeter.

CPC (Calling Party Control) A brief dc signal generated by your local central office when a caller hangs up.

CPU (Central Processing Unit) Also called a microprocessor. A complex programmable logic device that performs various logical operations and calculations based on predetermined program instructions.

cradle An area on a telephone's housing where the handset or portable unit may be kept when not in use.

DAC (Digital-to-Analog Converter) An electronic device used to convert a pattern of digital information into a corresponding analog voltage.

data In telephone systems, any information other than human speech.

decibel (dB) A unit of relative power or voltage expressed as a logarithmic ratio of two values.

demarcation point The point where a building connects with the outside wiring supported by the BOC. In a home, the demarcation point would be at the network interface connector.

demodulation The process of extracting useful information or speech from a modulated carrier signal.

diode A two-terminal electronic device used to conduct current in one direction only.

drain One of three electrodes on a MOS transistor.

DTMF (Dual-Tone Multi-Frequency) A process of dialing that uses unique sets of audible tones to represent the desired digit.

emitter One of three electrodes on a bipolar transistor.

EPROM (Electrically Programmable Read-Only Memory) An advanced type of ROM that can be erased and reused many times.

Exchange area A territory in which telephone service is provided without extra charge. Also called the local calling area.

FM (Frequency Modulation) A technique of modulating a carrier sinusoid with information for transmission.

full-duplex A circuit that carries information in both directions simultaneously.

gate One of three electrodes on a MOS transistor.

ground start A method of signalling between a telephone and the central office where a signal line is grounded to request service.

half-duplex A circuit that carries information in both directions, but in only one direction at a time.

harmonics Multiples of some intended frequency, usually created unintentionally when a frequency is first generated.

hybrid Also known as an induction coil. A specialized type of transformer used in classical telephones to couple the two-wire telephone line to an individual transmitter and receiver.

ICM (Incoming Message) The message that is left by a caller on an answering machine.

IF (Intermediate Frequency) A high frequency signal used in the process of RF demodulation.

impedance A measure of a circuit's resistance to an ac signal, usually measured in ohms or kilohms.

inductance The measure of a device's ability to store a magnetic charge, measured in henries, millihenries, or microhenries.

inductor A device used to store a magnetic charge.

LCD (Liquid Crystal Display) A type of display using electric fields to excite areas of liquid crystal material.

LED (Light-Emitting Diode) A specialized type of diode that emits light when current is passed through it in the proper direction.

loop current The amount of current flowing in the local loop.

loop start The typical method of signalling an offhook or line seizure condition where current flow in the loop indicates a request for service.

local loop The complete wiring circuit from a central office to an individual telephone.

modulation The systematic changing of the characteristics of an electronic signal in which a second signal is employed to convey useful information.

MTS (Message Telephone Service) The official name for long-distance or toll service.

NAM (Number Assignment Module) An erasable memory IC programmed with an assigned telephone number and specific identification information, typically used with cellular telephone circuits.

OGM (Outgoing Message) The message that a caller hears when an answering machine picks up the telephone line.

permeable The ability of a material to become magnetized.

piezoelectric The property of certain materials to vibrate when voltage is applied to them.

pps (Pulses Per Second) The rate at which rotary or pulse interruptions are generated. A rate of 10 pps is typical.

program A sequence of fixed instructions used to operate a CPU.

PSTN (Public Switched Telephone Network) A general term for the standard telephone network in the United States. The term refers to all types of wiring and facilities.

pulse A process of dialing using an IC (instead of a mechanical device) to generate circuit interruptions corresponding to the desired digits.

RAM (Random-Access Memory) A temporary memory device used to store digital information.

RC (Regional Center) Telephone facilities that interconnect both toll centers and some central offices, and support long-distance telephone service.

rectification The process of converting dual-polarity signals to a single polarity.

regulator   An electronic device used to control the output voltage or current of a circuit, usually of a power supply.

resistance   The measure of a device's ability to limit electrical current, measured in ohms, kilohms, or megaohms.

resistor   A device used to limit the flow of electrical current.

ring   An alerting signal sent from a central office to a telephone or other receiving equipment, such as an answering machine.

RF (Radio Frequency)   A broad category of frequencies in the range above human hearing, but below the spectrum of light, typically from 100 kHz to over 1 GHz.

ring   One of the two main wires in a local loop. The name originally referred to the ring portion of a phono plug that operators used to complete connections manually.

ROM (Read-Only Memory)   A permanent memory device used to store digital information.

rotary   A process of dialing that uses a mechanical device to open and close a set of contacts in a pattern corresponding to a desired digit.

sidetone   A small portion of transmitted speech that is passed to the receiver. It allows a speaker to hear their own voice and gauge how loudly to speak.

SMT (Surface Mount Technology)   The technique of PC board fabrication using components that are mounted directly to the surface of a PC board instead of inserting them through holes in the board.

SOT (Small-Outline Transistor)   A transistor designed for use with surface mount PC boards.

source   One of three electrodes on a MOS transistor.

subscriber loop   Another term for the local loop.

TC (Toll Center)   Telephone facilities that interconnect central offices.

tip   One of the two main wires in a local loop. The name originally referred to the tip of a phono plug that operators used to complete connections manually.

transistor   A three-terminal electronic device whose output signal is proportional to its input signal. A transistor can act as an amplifier or a switch.

transformer   A device using inductors to alter ac voltage and ac current levels, or to isolate one ac circuit from another.

VOX (Voice-Operated Control Actuation)   A circuit, usually used in answering machines, that detects the presence of a caller's voice and allows the machine to continue recording.

Bibliography

AT&T Technologies. 1977. *Telecommunications Transmission Engineering*. Indiana: AT&T.

Bernard, Joseph. 1986. *A Guide to Cellular Telephones*. California: Quantum Publishing.

Bigelow, Stephen. J. (March, 1991) *All About Cellular Telephones*. Popular Electronics: 58-63, 106.

_____. (March, 1990) *Telephone Answering Machines (Part I)*. Modern Electronics: 18-24, 72, 75-76.

_____. (April, 1990) *Telephone Answering Machines (Conclusion)*. Modern Electronics: 46-49.

_____. (August, 1988) *Telephone Technology (Part I)*. Modern Electronics: 18-23, 78, 79.

_____. (September, 1988) *Telephone Technology (Conclusion)*. Modern Electronics: 28-31, 98.

_____. 1991 *Understanding Telephone Electronics, 3rd edition*. Indiana: Hayden Books.

Greenfield, Joseph. D. 1983. *Practical Digital Design Using ICs*. New York: John Wiley & Sons,

Hardy, James. K. 1986. *Electronic Communications Technology*. New Jersey: Prentice-Hall.

Pasahow, Edward. J. *Microprocessors and Microcomputers for Electronics Technicians*. New York: McGraw-Hill, 1981.

Schuller, Charles. A. 1979. *Electronics: Principles and Applications*. New York: McGraw-Hill.

Index

Other Bestsellers of Related Interest

Ready-To-Build Telephone Enhancements
Delton T. Horn
Let best-selling author Delton Horn show you how to turn an average, everyday telephone into something much more . . . a fully functional, multi-purpose home communications system that will save you time, money and worries. It's a virtual feast of novel ideas for electronics hobbyists and do-it-yourselfers, including: a hold button with music, call timer with automatic hang-up, long distance lockouts, auto-answering modem, and ear-piece amplifier and flashing-ring projects for the hearing impaired.
ISBN 0-07-030412-2 $17.95 Paper

Old-Time Telephones: Restoration and Repair
Ralph O. Meyer
For all electronics hobbyists, telecommunication enthusiasts, and others who are just plain charmed by antique telephones, this is a unique guide to the history , technology, restoration and repair of telephones and telephone systems of the past 120 years. Collectors will find the antique telephone schematics and photographs worth far more than the price of the book. A definitive reference unlike any other available.
ISBN 0-07-041818-7 $19.95 Paper
ISBN 0-07-041817-9 $34.95 Hard

Modems Made Easy: The Basics and Beyond—2nd Edition
David Hakala
A practical guide to modems and the online world—for beginners and experienced users. Revised and updated to cover the latest developments in modem technology, including speeds up to 28.8K BPS and PCMCIA cards, Modems Made Easy gives you information about how to buy the model that's right for you and how to set it up, fine tune it, and troubleshoot problems. The book also covers doing business online, finding the best shareware, BBS's, online databases, software libraries, and free connections to the Internet.
ISBN 0-07-882116-9 $21.95 Paper

TV Repair for Beginners—4th edition
George Zwick and Homer L. Davidson
With this handy guide, TV owners can save themselves hundreds of dollars in repairs by repairing many of the most common TV problems themselves. This updated edition features in-depth coverage of the latest solid-state TV circuitry, high-definition TV, large-screen TV's and TV stereo systems. Features a wealth of schematics and illustrations that will help you keep your TV set operating like new.
ISBN 0-07-073092-X $19.95 Paper

Troubleshooting and Repairing Consumer Electronics Without a Schematic

Homer L. Davidson

An indispensable, hands-on guide for electronics technicians, students, and advanced hobbyists with hundreds of illustrations and photographs demonstrating how to locate, test, and repair defective components in: car radios, cassette decks, CD players, VCR's, and more, even when you don't have the schematic. For each type unit, Davidson includes a handy list of common mechanical symptoms, a time-saving troubleshooting chart telling where to look for specific problems, and actual case histories that explain how various repairs were made.

ISBN 0-07-015650-6 $22.95 Paper
ISBN 0-07-015649-2 $34.95 Hard

The Illustrated Dictionary of Electronics—6th Edition

Stan Gibilisco

No matter what the level of skill—student, hobbyist, engineer, or technician—this bestselling dictionary covers the entire spectrum of informational needs. Gibilisco thoroughly references the terminology of computers, robotics, lasers, TV radio, IC technology, digital and analog electronics, audio and video, power supplies, and fiberoptic communications. Terms are defined clearly and precisely, with as little technical jargon as possible, so that even beginning experimenters will understand them.

ISBN 0-07-023599-6 $32.95 Paper

Troubleshooting and Repairing PC Drives and Memory Systems

Stephen J. Bigelow

Stephen Bigelow gives reliable, up-to-date troubleshooting and repair guidelines for all types of PC storage and memory devices. He shows how to repair or replace hard drives, backup drives, RAM IC's, SRAM and DRAM chips, floppy drives, CD-ROM's, expansion SIMMs, and FLASH cards. Includes consolidated troubleshooting charts, vendor listings, a glossary, and a schematic symbols chart. Covers new tools, test equipment, and a diagnostic software, too.

ISBN 0-07-005314-6 $24.95 Paper

How to Order

Call 1-800-822-8158
24 hours a day,
7 days a week
in U.S. and Canada

Mail this coupon to:
McGraw-Hill, Inc.
P.O. Box 182067
Columbus, OH 43218-2607

Fax your order to:
614-759-3644

EMAIL
70007.1531@COMPUSERVE.COM
COMPUSERVE: GO MH

Shipping and Handling Charges

Order Amount	Within U.S.	Outside U.S.
Less than $15	$3.50	$5.50
$15.00 - $24.99	$4.00	$6.00
$25.00 - $49.99	$5.00	$7.00
$50.00 - $74.49	$6.00	$8.00
$75.00 - and up	$7.00	$9.00

EASY ORDER FORM— SATISFACTION GUARANTEED

Ship to:

Name

Address

City/State/Zip

Daytime Telephone No.

Thank you for your order!

ITEM NO.	QUANTITY	AMT.

Method of Payment:

☐ Check or money order enclosed (payable to McGraw-Hill)

☐ DISCOVER   ☐ AMERICAN EXPRESS Cards

☐ VISA   ☐ MasterCard

Shipping & Handling charge from chart below	
Subtotal	
Please add applicable state & local sales tax	
TOTAL	

Account No. ☐☐☐☐☐☐☐☐☐☐☐☐☐☐☐☐

Signature _____ Exp. Date _____
Order invalid without signature

In a hurry? Call 1-800-822-8158 anytime, day or night, or visit your local bookstore.

Key = BC95ZZA